WIRED WOMEN

· ·

GENDER AND NEW REALITIES

IN CYBERSPACE

EDITED BY

LYNN CHERNY AND

ELIZABETH REBA WEISE

SEAL PRESS

Library of Congress Cataloging-in Publication Data
Wired women : gender and new realities in cyberspace / edited by Lynn Cherny and Elizabeth Reba Weise.
Includes bibliographical references and index.
1. Women's studies—Computer network resources.
2. Women—Computer network resources. 3. Feminism—Computer network resources. I. Cherny, Lynn. II. Weise, Elizabeth Reba, 1962–
HQ1180.W57 1996   305.4'0285'467–dc20   95-51742
ISBN 1-878067-73-7

Printed in the United States of America
First printing, April 1996
10 9 8 7 6 5 4 3 2 1

Distributed to the trade by Publishers Group West
In Canada: Publishers Group West Canada, Toronto, Canada
In Europe and the U.K.: Airlift Book Company, London, England

Cover design by Kate Thompson
Text design by Clare Conrad
Composition by Partners in Design

Acknowledgments

from Beth

This book began as a great idea I had for someone else to do. I suggested it to Barbara Wilson of Seal Press back in 1993, when I was still getting over the exhaustion of editing my first book for the press.

Barbara said "Let's do it!" to which I replied, "Are you crazy? I'd like to get back to having a life, thank you very much."

But as I became more and more enmeshed in and enamored of this thing we call the online world, through both my personal and professional life, the need to hear women's voices and perspectives about cyberspace became overwhelming. Every time I saw Barbara I brought up the idea of an essay collection again, asking if she hadn't found someone to do it yet. The answer was always no.

Then in 1993 I met Lynn Cherny and we immediately hit it off over the twin topics of science fiction and the Internet. We'd been friends a few months when I first broached the topic of her possibly working on a book with me. Amazingly, given that she was deep in the middle of a demanding Ph.D. program at Stanford, she said yes.

I called Barbara and told her we'd do it. Lynn's email address became as familiar to me as my own and we plunged in. All too soon afterward, Barbara left Seal to work full-time on her own writing, a necessary move for her though sad for us.

One of the things which became clear during the course of this book is that the nascent old-girls network we used to talk about in our political discussion group days in Seattle has come to full flower. In Barbara's place as editor we got the indomitable June Thomas, a friend from my long-ago days writing for *off our backs* in the early 1980s and a woman whose enthusiasm for things online made her the perfect foil for us.

And imagine my pleasure when it turned out that Kate Thompson, someone I'd known from collective households of fifteen years ago, turned out not only to be the former art director of the *Village Voice*, but also the designer of our cutting edge cover.

Thanks are due in no small part to the authors, all fifteen of them, whose breadth of experience and professional attitude made this book fascinating to edit. I am honored to have been midwife to their creations.

Once again I had the pleasure of working with copy editor Cathy Johnson, whose careful reading of a text full of jargon and slang, most of which has yet to be crystallized into written form, resulted in a clarity that was beyond me. Friend and fellow Swediphile Anita Anderson graciously lent her eagle (and professional) eye to the manuscript in galley form.

And of course thanks to Lynn, who undertook this project during a moment of calm (and perhaps insanity) and stayed steady even when it became clear that the final edits would happen the same month her thesis was due. Her technical expertise, wide contacts and dry wit made it all a pleasure.

and from Lynn

Despite being overwhelmed by a dissertation, job hunting and chronic tendonitis (RSI from typing on bad equipment with no preventive education, thanks to Stanford University!), I found this book a welcome break from a sometimes too-dry academic scene.

Because my time was so limited, I owe huge debts to my co-editor who pulled up my slack when panic set in; her experience with Seal Press was an invaluable resource to draw on. I wouldn't have done it without her.

I also feel humbly grateful to the women I coerced to write for us, who dealt with an unreasonable amount of uncertainty and fumbling from me as a naive editor.

June Thomas, our editor at Seal, was a great support and calming influence when I was trying to move across the country and deal with last-minute issues. She's also a terrific dancer and one of the funniest people I have ever met.

All in all, this book made me new friends and expanded my horizons; I hope it does the same for our readers.

Contents

A Thousand Aunts with Modems

Elizabeth Reba Weise

I got online so I could break up with my girlfriend.

After several years in the trenches of reporting, I'd been offered a career-making job in California. The papers at home weren't hiring and if I wanted to stay a journalist, taking the job was the only option. Thirteen days before my lover and I were supposed to move, she sat me down on the couch and told me she wasn't going. I had my life to pack up, my friends to take leave of. There was no time to do anything except cry and divide up the kitchenware.

We both knew there were months of discussion needed before we would be through with each other. The emotional half-life of our relationship was glowing like uranium tailings in the living room. In my grief I could only put my books in boxes and say, "We'll deal with it when I get down there."

All this talking we needed to do, and no way to do it. My partner hated the telephone, she lost the nuances of emotion without a body to watch—felt adrift amid silences she could not pull meaning from. We had courted by letter, two writers knitting a relationship by post, crafting noun and adjective to make clear our precise meanings—words on a page were a natural medium for us. When we were falling in love the measured pace of the mail had only intensified our longing for each other. But in the frenzy of my need to understand our parting, I knew I could not bear the wait of a slow week for the questions and answers of letters, three days to Seattle, three days back to San Francisco.

Email, a medium I'd only read about, was the only way I could think to do it.

From the moment I had first used a computer for word processing in 1986 on an old "lugable" with a screen the size of a paperback, I was enamored of the enormous freedom the endless page offered. A horrible typist and worse speller, the computer offered me unheard of freedom. I longed for the malleability of text made possible by the blinking cursor. But the world I roamed there was interior only. I had no modem, no electronic mail. My words lived on disk, and when I wanted someone to read what I had written, I handed them paper and ink. Nothing else had been necessary.

My first home computer came to me in 1988 through an under-the-table temp job at a local computer dealer. For two months I spent the days I wasn't doing freelance radio reporting for National Public Radio (NPR) working for barter in the dark basement office in my employer's house. At the end of that time I went home with a demonstration PC, with a 286 and a 5 1/4-inch floppy disk drive and no hard drive at all—close to state of the art at the time, and, in my world, wildly expensive. I didn't know one other person who had a computer at home.

That night I plugged keyboard to computer to screen and sat down to write. Words poured out of me. My computer felt to me, feels to me still, the way wings must feel to a bird, making flight possible.

For years that solitary discipline before the cursor was the extent of my use of computer technology. My computer evolved slowly, based only upon what I really needed it to do. When I signed the contract for my first book, my sister came over and helped me install a hard drive, all 30 mega-bytes of it. We sat on the floor of my room with the guts spill-ing out of the box, one of my housemates kibitzing from the living room. I'd done some reporting about computer issues for NPR, but my sister was the only one in that room who had any idea what she was doing.

Three years later it was that same gray box, my old 286 with its souped-up 30-meg hard drive, that my ex opened up as we took leave of each other the best way we knew how. She cursed and figured as she pulled the modem from her computer, the one she'd done almost all of her doctoral work on, and put it into mine. The shareware she copied over to my hard drive was as clunky and basic as anything written, but it worked for text, which was all I cared about.

In my circle, computers aren't fetish objects, but tools. They don't get thrown away, but are endlessly recycled. No one cares about having the latest or the fastest system. Just having something to type on that isn't a pay-by-the-hour computer at Kinko's is a big deal.

My ex gave me her modem because she had access to email at work. When she could afford a laptop to do her field-work with, she gave her old system, monochrome and slow, to a writer friend who until that point had to stay late at work typing out his stories from notebooks he kept longhand. It's still on his desk, 9-pin dot-matrix printer and all. When I finally had to buy a Windows-capable 80486 to keep up with the technology I was writing about, I carted my beloved gray box off to a storefront mailing place and shipped the whole thing back up to my sister, who now uses it to write her papers for another bout of school.

It was that old, slow, system that came with me on the two-day drive down the coast. But despite its new modem, it was mute. My first week in San Francisco a friend walked down the hill from her house to mine to show me how to make the words flow.

I dragged the computer out from my little study and put it on the dining room table together with a phone cord stretch-ed from the kitchen and a light borrowed from the living room. Stephanie and her partner, Bill, showed me how to find my modem, how to tell it to make a phone call, how to connect up to a new world. When they left that night, I could call forth a new, nonblinking cursor on my screen, a cursor that was no longer the door to my own solitary world, but to the outside.

I walked through it.

For months I had lived a two-dimensional life, going to work and then coming home to pour out my feelings in letters that appeared on my ex's screen minutes after I sent them. Our conversations ran to tens of thousands of words. Slowly, all these words flung back and forth through the ether leached off the worst of the pain, and we made a kind of peace. Finally I was able to raise my head out of the fog I'd been living in and look around to see where I was.

Physically that took the form of venturing out into my new city of San Francisco. But in the months in which my ex and I had been teasing apart our hearts online, I'd slowly realized that I'd taken up residence in another place—a small town at the edge of cyberspace called the WELL.

Stephanie and Bill had shown me how to connect to the Internet service provider they knew best, the one where they had their own accounts. All I knew at the time was that I needed those little words after the @ sign so my letters could find their way to the woman I was learning not to love.

Without the vocabulary or grammar of the language of this new world, I did what all beginners do, and wrote down verbatim each word Stephanie and Bill had typed on the screen. Later that night I parroted them back on my own and was astounded when the genie answered and the then-revelatory words "Welcome to the WELL" appeared on my screen, words which for the first time came not from inside the gray box that sat beside my keyboard, but from someone else's.

Just as I began to come back to life, the dues-paying portion of my new position at The Associated Press began and I was assigned the graveyard shift. I went to work at eleven at night, the bureau emptied of reporters by midnight and I was alone filing the wire until six in the morning when the news editor arrived to begin the new day.

The second night I was on the shift, I backed out of the AP's utilitarian editing system and reconfigured the communications software so I could dial in to the WELL from my terminal. It took me all of that dinner hour and a fair amount of my dinners over the next few nights to get it to work, but finally I could sit in front of that glowing screen at three in the morning and, while the city slept around me, enter a bustling world that never closed.

I was ready now for something beyond letters coming from Seattle. Night after night alone in the bureau, I would throw my dinner in the microwave and then come back to eat at my "tube." Tupperware in one hand, the keyboard under the other, the police scanner muttering in the background, I slowly found connection.

The WELL gives full access to the Internet, but mostly it is a place unto itself. It consists of hundreds of discussion areas, called conferences, on topics as widely varied as parenting, the Grateful Dead and the First Amendment. Each conference is a group of people having several ongoing conversations, a written-down dinner party where the guests are loud and opinionated, thoughtful and, sometimes, hysterical. I read the Media conference, and News, and Books and Writers. I poked my head in on Gen-X and Star Trek. But it was in WOW—Women on the Well—that I found a home.

WOW is a private conference, open only to women. I had to call one of the hosts to confirm that I was indeed female. She then put me on the user list so that when I typed the words "go WOW" at the prompt, I was admitted into this community of women.

I slept days, worked nights, became so pale as to look sickly and barely saw another soul, but each night at the bureau, surrounded by empty desks and the velvet darkness outside, I logged in to WOW to read the stories of these women's daily lives. Their words, written perhaps in the middle of the day, were waiting for me like a letter on the kitchen table.

I came in to the community with a cry of pain, feeling alone and bereft, and these women I did not know sat down beside me and offered comfort, told their own stories of break-ups and partings, of finding their way in a new city, of making a life where you did not have one before. Every night while I ate my dinner under the fluorescent lights, I would read the stories of their daily triumphs and frustrations—a traffic ticket, a new pair of pants that actually fit, boyfriends, girlfriends, husbands and children, what to do about bullies at school, how a leach field for a septic tank really works.

The hum of the wire room printers faded away as I fell into their stories. Someone got a job, we all congratulated her, post after post, a receiving line of good wishes. Someone else lost her job, and women from as far away as Tokyo offered their support and résumé-writing tips.

I learned a lot in my late-night sojourns to the WELL. I who had few heterosexual friends met happily married women. I listened in as women with experiences vastly different from mine discussed blending families, tax law, how much to tip in a hotel, what it had meant to come of age in the sixties and how to cope when one's parents died.

In a way, the WELL, and WOW in particular, was like being given the gift of an extended family, something I had never experienced. Suddenly every night I had family dinner, could sit back and nibble at my food while aunts and uncles and cousins argued and told stories about people I had never met, but whose experiences enriched me.

I remember one woman who wrote about letting her gifted son go off to a boarding school where she knew he would finally find peers. She told us the story of the day he was born, when she held him and said, "Okay, I know the whole object of this game is to let you go, and I really will." I felt honored to be allowed into her life, into the lives of all the women of WOW. Alone in a new city, I felt surrounded by a thousand aunts—a thousand aunts with modems.

Another kind of learning was taking place as well, a real-time, real-world, real-friend kind of learning. I've got a good thirty computer books on the shelf behind my desk, but the true leaps in understanding have taken place when someone sat down beside me and walked me through something new, calling out words for me to type and explaining what was appearing on the screen in response.

By winter the language of this new place, the WELL, fit my tongue, the words came easily and I found I wanted the challenge of new dialects. About that time some friends in Oakland, Gayle and Richard, invited me to a dinner to introduce me to a friend of theirs they thought I'd like, someone with whom I shared a common interest in linguistics and science fiction.

They were right, and a few weeks later I was sitting in

front of my computer, this time next to a new friend with years of experience on the net who calmly and without jargon explained the intricacies of reading Usenet newsgroups. I'd never even stepped out into the Internet before, and suddenly Lynn Cherny—who was to become co-editor of this book—had me gliding through the baroque fullness of newsgroups, touching down to check out points of interest.

Not long after that, I made my way one chilly January morning to a community seminar about the online world. Some sixty people crowded into a musty performance space in the artsy South of Market part of town to hear two guys from Bay Area Internet Literacy race through the basics of getting online. Looking around the room at a crowd made up of community activists, artists and grandmothers listening eagerly as the presenters waxed poetic about modems and gopher spaces and bulletin boards, I knew this stuff wasn't just for computer geeks.

Both the guys teaching the workshop had accounts on the WELL, and one of them, Eric Theise, became a friend. He came by and showed me how to use Internet applications such as gopher and archie, and together we figured out how to get my World Wide Web connection working. As I got more into the online world and began writing about it for the AP, Eric spent many hours on the phone giving me, in patient detail, the technical background I needed.

I will admit that I have been very lucky in the connections and community I have found online. Not everyone's experience is so uplifting. But perhaps part of that luck comes from not being afraid to ask for help, and the presumption that reaching out to others will teach me more than trying to always figure it out on my own—a trait which is perhaps more common among women then men.

The day my hard drive crashed because of a computer virus from a disk I'd brought from work, I spent a few hours fussing, then calling in help. One friend came and she called a friend of hers and over the course of a long evening, interrupted by a run to the store for more chocolate, the three of us were able to partially reconstruct it. We don't quilt, but we do defragment hard drives while gossiping furiously.

I have learned an incredible amount from my friends, from Eric, from Lynn, from WOW's technical questions discussion group. My sister talked to me about hardware;

Richard sent me lovingly detailed spec lists to use when I went computer shopping. It was a communal effort: when one person learned something new, it was quickly scattered through lines of friendship until everyone had the skill, or at least knew where to get it.

In the past two years, Lynn has finished her Ph.D. with a thesis on computer-mediated communication on Multi-User Domains (MUDs) and gotten a job at Bell Labs, sort of like graduating from journalism school and getting your first job at the *New York Times*. I started writing a weekly column about the Internet for The Associated Press, with the goal of explaining what all the hype was about in a way that wasn't "techier-than-thou." In November 1995 I was made AP's national cyberspace writer.

This book, like my initiation into cyberspace, came out of a web of connection so dense you could walk across it. I announced the project on the WELL, and lists of names of possible contributors poured in, along with a note from Stephanie Brail saying she was interested in writing about her experiences with online harassment. Eric Theise, off the top of his head, came up with a page full of women with plenty to say. On that list were Karen Coyle, a brilliant speaker, and the wonderfully snarky Paulina Borsook. Paulina had an engineer friend, Ellen Ullman, who she thought I should meet, and over lunch another essay was proposed. Eric organizes the Jacked In Cyberspace Literacy Series, and he asked Lynn, Karen and I to be on a panel about women online. Linguist Laurel Sutton came up after the talk and said she'd be interested in writing for us, and suddenly Emily Post had a place in the book.

Karen Coyle and I went to the Computers, Freedom and Privacy conference put on by Computer Professionals for Social Responsibility, and after hearing Jean Camp talk I asked her if she'd like to contribute to the book. Her officemate at Carnegie Mellon, Donna Riley, also turned out to have something to say about online censorship.

Lynn's work with MUDs, online worlds where people hang out and talk in real time, had introduced her to a lot of interesting women who worked and played in MUDs and other online communities. She asked MUD friends Tari Lin Fanderclai, Judy Anderson, Michele Evard, Shannon McRae and Lori Kendall if they'd be interested in writing about their experiences for the book. Susan Clerc, a friend of Lynn's from media fandom, was writing about the impact of the net on the women-dominated fan communities, and Lynn asked her to contribute as well.

Netta Gilboa appeared in an issue of the *Computer Underground Digest* and piqued our interest with her perspective on the hacker underground. Lynn heard Judy Malloy speak about her experiences writing hypertext fiction and she and her hypertext co-author Cathy Marshall were added to our growing list of authors.

Our efforts were lacking in two areas, unfortunately. None of the few women of color we were able to find online were available to write, a mirror of the extremely white nature of the medium at this time. Nor were we able to find anyone to write about the specific experiences of women on the large commercial services such as America Online, CompuServe, Delphi and the Microsoft Network, most likely a result of our own Internet-based bias.

Computer-mediated communication is not impossible to learn; it's not brain surgery. It is friends teaching friends, using what you need, ignoring what you don't, a tool like any other. Neither is it a replacement for reality. Don't imagine in all this furious typing that my other connections faded away.

The people I have met through cyberspace came to me through human connections. If they are local, I almost always choose to see them face-to-face rather than through words on the screen. Email will never be a replacement for talking. I still speak weekly, sometimes daily, with friends in Seattle and elsewhere. Telephone bills are the modern price of exile. I still give dinner parties in real life and love sitting around a table talking until 2:00 a.m., unable to leave because the conversation is too good. These things remain.

Being online is an adjunct, a backyard fence, a coffee shop, a favorite hangout, a weekly support group. It's not my life, but it's a nice medium to have in one's life. It is not a social revolution, but at times—in my case, at one very specific time—it can be a revelation.

REFLECTIONS ON A NEW WORLD

Come In, CQ:
The Body on the Wire
. . . .
                       .
Ellen Ullman   .
                       .

There is a male sort of loneliness that adheres in programming. It's nothing like women's loneliness, which might be
assuaged by visits and talk and telephone calls, an interrupting sort of interaction that might come anytime: while you're
cooking dinner, or dressing, or about to leave the house. Programmer loneliness does not interrupt. The need for con-
centration forbids it. If there must be "talk," it must be of the
ordered, my-turn, your-turn variety—asynchronous, sent and
stored until the recipient decides to check his email.

There's no substance to this email, of course, no rattle in
the doorslot or clatter to the floor. Even responses are rare.
Programmers reply by exception: You'll hear soon enough
about errors, arguments and disagreements. But all other
possible replies—they agree, they don't care, they're home-
sick, they're not reading mail today—all that is signified by
silence. Fifteen years of programming, and I'm used to the

. . . .

silence. I've become accustomed to the small companion-ships of clicking keys, whirring fans and white noise. Fifteen years of programming, and I've finally learned to take my loneliness like a man.

When I was growing up, the boy next door was a ham radio operator. His name was Eugene. He was fat, went to Bronx High School of Science to study engineering, and sat evenings in the basement of his house beaming a signal off into the atmosphere. The heart of Eugene's world was the radio room: a dim box filled with equipment, all of it furnished with dials and toggles and switches. It was there he spent his Saturday nights, alone in the dark, lit only by small red lights and a flex-arm lamp bent low over his operator's guide.

I grew up in the shadow of Eugene's radio. Over time, his antenna become more and more elaborate, and my family kept giving him permission to add anchors to the roof of our house. From a simple T-bar arrangement, the antenna sprouted new masts and crossbeams, and finally a wide circular thing that could be positioned with a motor. This whole complicated structure whirred when the motor was engaged, vibrated in the wind, was twice reduced to dangling pieces by hurricanes. Mostly, it just sat there and cast an electronic shadow over our house, which is how I came to know everything about Eugene's secret life in the basement.

On Saturday nights, when my parents and sister were out, I could hear Eugene and "see" him on the wire. Perry Como would be singing on the TV set, and then, suddenly, the loud white noise of electronic snow. Through the snow came a pattern like the oscilloscope on *Outer Limits*, which I came to think of as the true physical presence of Eugene, the real Eugene, the one he was meant to be beyond his given body. He always seemed to be broadcasting the same message: "CQ, CQ. Come in, CQ. This is K3URS calling CQ. Come in, CQ." K3URS were his call letters, his license number, his handle. CQ meant anyone. *Come in, CQ*: Anyone out there, anyone at all, if you're there, please respond. To this day, nothing reminds me of engineering loneliness so much as that voice calling CQ through the snow.

Sometimes Eugene actually made contact. Breaking through the television signal came both sides of their "conversation." What they did, it seemed, was compare radios.

All those massive structures rising over neighborhoods, all that searching the night sky for another soul on the air, and then they talked about—equipment. One talked—my amp, my mike, over; then the other—my filter, my voltage regulator, over. This "talk" seemed to make them happy. I could hear them laughing: a particularly wide pattern of amplitude, a roiling wave across the screen. If CQ was the representation of loneliness, then this pattern was the look of engineering fulfillment. It reassured the boys in the basement: All that hardware has a purpose, it said. It can indeed bring you company.

Thirty-five years later, I have insomnia, but down the hall my three computers are sleeping. Not sure what I'm looking for, I go wake them up. The Mac PowerBook is really sleeping: Some hours ago, I put it in "sleep mode," and now its small green light is blinking as steadily as a baby's breathing. The portable Sun workstation, Voyager, shows a blank screen. But the touch of a key puts it right back where I left off five hours ago. One small window opens to show a clock. I know the clock is digital, but for some reason, I'm glad it's been given a face, a big hand and a little hand and a second-hand sweep, all of which now say it's 2:05 a.m. PST. The last machine, the PC, is primitive. It doesn't really know how to go to sleep. Like a cranky child, it needed diversions and tactics to be put down for the night: a screen saver that knows when to come on, a human who remembers to hit the right off buttons.

The room is filled with the sound of fans and disk drives spinning to life. Two big 21-inch monitors give off a flickering light. Still, flicker and all, I admit I'm happy. I *like* sitting in a humming room surrounded by fine machinery. I dial up my three Internet accounts one after the other. The net is full of jabber and postings from around the globe. But now I know what I'm looking for, and it's not there. I'd like to find someone still up and working on a program, someone I know—a colleague on my node or one nearby, who'll get my mail virtually "now."

Sometimes I do find someone. Although almost no one answers mail in real time during the day, a kind of license prevails in the middle of the night. "What are you doing on at

2 a.m.?" the colleague writes, finding my mail when the signal he's set on his machine beeps to say there's "incoming." He knows, but here, online at 2 a.m. one does not say *I'm alone, I'm awake, Come in, CQ*. What am I doing on in the middle of the night? I know his workstation has the same small window holding a clock with a face. "Same as you," I reply.

The next morning we see each other at a meeting. We don't mention we've met in the middle of the night. Daytime rules prevail: We're about to have a no-rules battle over a design issue. We can't possibly think about the person who was lonely and looking for company. That life, the one where our insomniac selves met, exists in a separate universe from this one, here in this room, where we're sitting next to each other at a conference table and about to do technical battle. Some implosion may occur, some *Star Trek*–like breach of containment fields may happen, if the two universes meet. No, the persona online must not touch the person at the table. As the meeting starts, I'm distracted. I want to ask him, "How are you? Did you get some rest?" He's inches from me, but in what way am I permitted to *know* him? And which set of us is the more real: the sleepless ones online, or these bodies in the daylight, tired, primed for a mind-fight?

Somehow, in the thirty-five years between Eugene's ham radio and my middle-of-the-night email, the search for electronic companionship has become a sexy idea. I'm not sure how this happened. One year I found myself exchanging messages with a universe of Eugenes, and the next, journalists were calling me up and asking if I would be an informant for a "phenomenological study of email."

This craze for the Internet, it's become a frenzy because of the Web. The pretty point-and-click navigators. The pictures and sound. The Rolling Stones' live broadcast. The Web is turning the net into television—TV for the ostensibly intelligent. It may not be acceptable to say that you have been up all night roaming through the high, weird channels on the cable. But somehow it's fine, impressive even, to say that you clicked around for ten hours on the Web.

The Web has a pretty face. But, underneath the Web is, well, a web. Of FTP sites. IP addresses. Tar files.[1] In this tangle of machinery, email crosses technical boundaries, significant bit orders are properly rearranged, parity bits get adjusted. It's all there to see in the email header.

```
From jim@janeway.Eng.Neo.COM Thu Apr 27 11:22:45 199
Return-Path: <jim@janeway.Eng.Neo.COM>
Received: from Neo.COM by netcom11.netcom.com (8.6.12/
Netcom) id KAA15536; Thu, 27 Apr 1995 10:55:59 -0700
Received: from Eng.Neo.COM (engmail2.Eng.Neo.COM) by
Neo.COM (komara.Neo.COM)id AA15711; Thu, 27 Apr 95
10:43:37 PDT
Received: from janeway.Eng.Neo.COM (janeway-20.Eng.Neo.COM)
by Eng.Neo.COM (5.x-5.3)id AA29170; Thu, 27 Apr 1995
10:42:06 -0700
Received: from hubris.Eng.Neo.COM by hubris.Eng.Neo.COM (5.0
-SVR4)id AA13690; Thu, 27 Apr 1995 10:42:05 +0800
Received: by hubris.Eng.Neo.COM (5.0-SVR4)id AA10391; Thu, 27
Apr 1995 10:42:04 +0800
From: jim@janeway.Eng.Neo.COM (Jim Marlin)
Message-Id: <9504271742.AA10391@hubris.Eng.Neo.COM>
Subject: Design notes due
To: dev-team@hubris.Eng.Neo.COM
Date: Thu, 27 Apr 1995 10:42:04 -0800 (PDT)
X-Mailer: ELM [version 2.4 PL21]
Content-Type: text
Status: R
```

This is the true face of the Internet. Most readers don't even look at the header, screening it out like static on a cordless phone. Yet the header holds the real path, machine to machine, the handoff of bits from system to system which takes place under the Web's pretty pictures and sound, under the friendly email windows of America Online and Prodigy. Without the covers, the Internet is still the same old fusty place created by the Department of Defense. And it retains its original motive: a place for the Eugenes of the world to exchange information about, say, rocket valves or caching algorithms. It's where the daily work of engineering takes place, in the famously arcane UNIX operating system, where the shortest possible command is always preferred.[2]

Although few managers would be likely to admit it, an engineer's place in the pecking order is largely determined by an electronic persona who lives in the interlocking email distribution lists called "group aliases." Every engineering project has its group alias—an Internet "address" that sends mail to all members of the team. Names come and go on the alias; people get "attached" and "unattached" with some regularity. Unless you're directly on the node where the alias is defined, or someone makes a point of telling you who's on now, you're never quite sure whom you're addressing.

Often, there are several aliases, names that include ever-widening circles of recipients, from the developers and project leads, to senior managers, to heads of other departments and so on out to the world. It's nearly impossible to know, at any given moment, who exactly is attached. On a recent project, one alias connected programmers and managers from California with managers in New Jersey; after that, other aliases disappeared into more distant time zones in Europe, Japan, India. Once, years ago, I slipped on the "To" line. Using the wrong alias, I inadvertently told a product manager just what I thought of his ideas. My colleagues in the development group—not reading the header, of course, and assuming from the content that we were "alone"—jumped right in with a fine round of character assassination. "Those who can't do become product managers," was the nicest thing said. An alias-slip only needs to happen to you once. Twice would be suicide.

For an engineer, gaining comfort and skill in using these various aliases—and creating the right online persona for each—is a prerequisite for surviving in the profession. Everything happens there: design, technical argument, news, professional visibility; in short, one's working life. Someone who can't survive by email has to find another way to earn a living. If an engineer begins to insist on too many meetings or too many phone calls (womanish, interrupting sort of interactions), he or she will soon be seen as a nuisance and a "bad programmer." Early in an engineer's life, one learns to send mail.

Life in the group alias is not an especially friendly place. Being on the project distribution list is akin to being the object of a Communist criticism/self-criticism session. Your colleagues have learned to exert technical influence by ferociously attacking your work while vehemently defending their own. It is a place purposely constructed to be a shooting gallery without apologies. What occurs there is a technical battle fought in the arena of technology—a tightening circle of machine reference. In a McLuhanesque way, cyberspace carries its own message back to the engineer: We are mind and machine mediated through mind and machine. A typical posting: "You are running in tautologies. Your whole way of thinking is a tautology. Or else you are stupid."

In this online battle, there is no sight of the victim's defensive posture, of course, no expression of fear and dismay; the wire gives off no smell of a human under attack. The object of attack must tough it out or quit. The sight of virtual blood on the screen is like running from a grizzly: It only makes the bear want to chase you. As one project leader put it, "We try to encourage arrogance."

The only recourse is humor. It is acceptable to designate oneself "the goat of the week." It is fine to say something like: "I agree to hold goatship for seven days, or until someone else commits an error of greater or equal stupidity, whichever comes first." But, under no circumstances, may anyone ask for compassion. For such sentiments, you must go to personal email, point-to-point, perhaps some middle-of-the-night search for company which must be refuted by day. No, you can't ask anyone to back off. The group alias is no place to look for love.

Is it any wonder then that engineers look for company on the net? While most of the world would think of Usenet conferences and Web pages as a degraded form of human communication (compared to, say, a dinner party or even a business phone call), for the average engineer the Internet represents an improvement on daily life in the group alias. The wider net—the conferences, the Web—offer release from the anxiety and claustrophobia of group email. They are places to find anonymity if one chooses, to be stupid (or arrogant) without consequence in "real" life. Travel to far-off places. Have fascinating discussions with erudite scientists around the world. Unburden yourself to a stranger. "Talk" without ever being interrupted. All this and more awaits the visitor to the wider net.

It has taken me a long time to understand why most women engineers I've known did not often fight their technical battles through the group alias (and why we therefore did not need the counterbalance of Internet conferences). We knew it was simply easier to walk down the hall to someone's office, close the door and have a talk. We "codeswitched"—changed modes of communication—as we found it necessary. We might take someone out to lunch, arrange a meeting, drop in for a chat or use the alias. Not all women can codeswitch—I've known some who never left their office; one bragged she had no interest whatsoever in physical existence and, as evidence, told us her home was not permitted to contain a single decorative object. But, being women as well as engineers, most of us can communicate on multiple channels. We use the Internet as a tool, like the phone or the fax, a way to transmit news and make appointments. For women, online messages constitute one means of communication among many, one type of relationship among many. Maybe this is why there are fewer of us online: We already have company. For the men, their online messages *are* their relationships. They seem content in the net's single channeledness, relations wrapped in the envelope of technology: one man, one wire.

There is, therefore, a usual gender-role reversal in the way men and women use the Internet. Men net-surf the way suburban women of the 1950s and 1960s used the telephone: as a way to break out of isolation. For nothing in today's world so much resembles the original suburbia as the modern software-engineering campus.

Close by a freeway on-ramp, meticulously planned and laid out, the engineering campus is nothing if not a physical and mental Levittown. It is endowed with artificial nature—bushes and hedges to soften the lines of parking lots. It reassures its inhabitants with splashing fountains, faux waterfalls, and fake lagoons where actual ducks sometimes take up residence. And there are those regular rows of offices: ranch houses for the intellect. Better ones overlook the lagoons; lesser ones, the parking lot. But, aside from small differences in size and orientation, the "houses" are all alike. The occupants are supposed to be comforted by the computerized equivalent of the washer-dryer and all-electric kitchen: workstations, network connections, teleconferencing cameras—*appliances.*

There, in this presumed paradise, engineers are stranded in the company of an infantile mentality: the machine.

The computer, as the engineer sees it, makes a toddler seem brilliant. For what engineers do is create artificial smartness. Our job is to make a simulacrum of intelligence, a thing that seems to contain knowledge only because it has been programmed to behave that away. We are the ones who create the pretty pictures and sound; we make the point-and-click interfaces. But the thing we talk to all day may be little more than a mechanism that reads bits off a disk drive. It does not "understand" us. If a comma is out of place, it complains like a toddler who won't tolerate a pea touching the mashed potatoes. And, exhausted though the programmer may be, the machine is like an uncanny child that never gets tired. This is the general definition of the modern software engineer: a man left alone all day with a cranky, illiterate thing, which he must somehow make grow up. It is an odd gender revenge.

Is it any surprise that these isolated men need relief, seek company, hook up to the net? Cyberspace: the latest form of phone yakking. Internet: mother's little helper for the male engineer.

This is not to say that women are not capable of engineering's malelike isolation. Until I became a programmer, I didn't thoroughly understand the usefulness of such isolation: the silence, the reduction of life to thought and form; for example, going off to a dark room to work on a program when relations with people get difficult. I'm perfectly capable of this isolation. I first noticed it during the visit of a particularly tiresome guest. All I could think was, there's that bug waiting for me, I really should go find that bug.

Women are supposed to prefer talking. I've been told that women have trouble as engineers because we'd rather relate to people than to machines. This is a thorough misconception. The fact that I can talk to people in no way obviates my desire (yes, *desire*) to handle a fine machine. I drive a fast car with a big engine. An old Leica camera—miracle of graceful glass and velvety metal—sits in my palm as if attached, part of me. I tried piloting a plane just to touch it: Taking the yoke into my hands and banking into a turn gave me the indescribable pleasure of holding a powerful machine while *it* held me. I'm an engineer for the same reason anyone is an engineer: a certain love for the intricate lives of things, a belief in a functional definition of reality. I do believe that the operational definition of a thing—how it *works*—is its most eloquent self-expression.

Ironically, those of us who most believe in physical, operational eloquence are the very ones most cut off from the body. To build the working thing that is a program, we perform "labor" that is sedentary to the point of near immobility, and we must give ourselves up almost entirely to language. Believers in the functional, nonverbal worth of things, we live in a world where waving one's arms accomplishes nothing, and where we must write, write, write in odd programming languages and email. Software engineering is an oxymoron: We are engineers, but we don't build anything in the physical sense of the word. We think. We type. It's all grammar.[3]

Cut off from real working things, we construct a substitute object: the program. We treat it as if it could be specified like machinery and assembled out of standard parts.[4] We say we "engineered" it; when we put the pieces of code together, we call it "a build." And, cut off from the real body, we construct a substitute body: ourselves online. We treat it as if it were our actual self, our real life. Over time, it does indeed become our life.

I fell in love by email. It was as intense as any other falling in love—no, more so. For this love happened in my substitute body, the one online, a body that stays up later, is more playful, more inclined to games of innuendo—all the stuff of romantic love.

I must stress from the outset there was nothing in this online attraction of "sexual harassment" or "environments hostile to women." Neither was it some anonymous, fetishistic Internet encounter. We knew each other. We'd worked on the same project off and on for years. But it was a project that took place almost entirely via Internet. Even the software was distributed through FTP sites; we "knew" our customers by their Internet addresses. I was separated from the development team by some fifty miles of crowded freeway, and I saw actual human beings perhaps once every two months. If I were going to fall in love on this project, there was no choice:

It would have to be by email.

I'll call the object of my affections "him" or "Karl," but these are only disguises. I'll describe coastlines and places that sound like San Francisco, but such descriptions may or may not be accurate. The only thing you can know for sure is that something like this did indeed happen, and that the "I" in the story is I, myself, contractor on the project.

The relationship began after a particularly vicious online battle. The thread went on for weeks, and the mail became progressively more bitter, heedless of feelings, sarcastic. My work was the object of scorn. I say "my work," but the team made no nice distinction between "me" and "my work." One wrote, "Wrong, wrong, wrong, wrong! Completely dumb!" Said another, "What's the objective? Just to produce some piece of shit to satisfy the contract?" If I hadn't been working around people like this for years, I surely would have quit. As it was, I said to myself, "Whoa. Remember they treat each other this way. It's just the far end of the scale on arrogance."

After I had been run through the gauntlet, Karl did this amazing thing: He posted to the group alias a story about the time he made a cut-and-paste error and therefore became "the official project whipping boy." He described how it felt to be the object of ridicule, and ended with the report of yet another stupid mistake he had just made. I watched this posting roll up my screen in amazement. In all my experience, no male engineer had ever posted such a letter to his colleagues.

To the group alias, I sent the following reply:

Thank you, Karl, for sharing the whipping energies with me. Your company at the post was much appreciated.

Even as I typed a period at the beginning of a clear line and hit the Return key—sending this mail off to the entire project group—I was aware of a faint whiff of exhibitionism. His reply only enhanced the thrill:

Delighted. Anytime.

Then we abandoned the group alias.

What followed were months of email that rode back and forth between us with increasing speed. Once a day, twice a day, hourly. It got so I had to set a clock to force myself to work uninterruptedly for an hour then—ring!—my reward was to check my mail. We described our lives, interests, favorite writers, past work projects and, finally, past lovers. Once we got to lovers, the deed was done. It was inevitable that we would have to go out, *see* each other. Yet we delayed. We wanted to stay where we were: in the overwhelming sensation of words, machine, imagination.

It's tempting to think of these email exchanges as just another epistolary romance—*The Sorrows of Young Werther* with phone lines. But the "mail" in electronic mail is just a linguistic artifact. Lasers can be described in terms of candle power, but there's no flicker, no slow hot drop of wax in laser light; and there's not much "mail" left in email. I have in my desk drawer a piece of paper on which Karl has written the title and author of a book. Here is his writing: precise and printlike, standing straight upward, as lean and spare as his body. Having this piece of paper, I know what the email lacks: the evidence of his flesh, the work of his *hand*.

And, although we seem to be delaying, prolonging the time of imagination, the email is rushing us. I read a message. The prompt then sits there, the cursor blinking. It's all waiting for me to type "r" for "reply." The whole system is designed for it, is pressing me, is sitting there pulsing, insisting, *Reply. Reply right now.* Before I know it, I've done it: I've typed "r." Immediately, the screen clears, a heading appears, "From:" my Internet address, "To:" Karl's address, "Re:" Karl's subject. And now I reply. Even though I meant to hold the message a while, even though I wanted to treat it as if it were indeed a letter—something to hold in my hand, read again, mull over—although my desire is to wait, I find it hard to resist that voice of the software urging, *Reply, reply now.*

There's a text editor available. I can fix mistakes, rethink a bit here and there. But there'll be no evidence of my changes, which makes edited email appear rather studied and, well, edited. No, the system wants a quick reply, and, according to some unspoken protocol, no one wants to look as if he or she had actually spent much time composing. So the ironic effect of the text editor is to discourage anyone from using it. It's best if the reply has the look of something fired off, full of spelling errors and typos. Dash it off, come to the beginning of a clean line, type a period, hit the Return key, and it's gone:

done, replied to.

What's missing now is geography. There's no delightful time of imagination as my letter crosses mountains and oceans. In the world of paper mail, now is when I should be hearing my own words in my lover's mind, envisioning the receipt of the envelope, the feeling at seeing the return address, the opening, the reading. But my email is already there. And my lover has the same pressures to type "r" as I did. Before I know it, it's back. "Re:" the same subject. Even though we're both done with the subject and haven't mentioned it for weeks, the subject heading lingers, back and forth, marker of where this thread of messages began.

Still, Karl and I do manage to forge a relationship out of this environment designed for information exchange. He meticulously types out passages from Borges, which we only admire, never analyze. We share a passion about punctuation. He sends me his dreams. I send him pieces of articles I'm working on. An electronic couple, a "we," begins to evolve: "We think that way," he writes once; "You and I feel that way," he says later. Suddenly, we change our signatures. He ends his messages with —K, I respond as —E, like adulterous correspondents who fear discovery.

But soon we come to the first communications problem of our relationship: interpolation. The email software we are using allows the recipient to copy the contents of the received message into the reply. At the beginning of an empty line, the recipient enters "~m" and the machine answers, "interpolating message number *nnn.*" The result is something like the following:

> There's something in this team's working process that's really broken. [I write in the original message]

I couldn't agree more.[Karl interpolates his reply]

> I think it's because they evaluate the messenger, not the ideas. I mean, when someone makes a suggestion, the immediate reaction is not to consider the idea but to decide if the person is worthy to be commenting on their work.

Interesting. I've felt alienated for a long time, but perhaps it takes an outsider to see exactly what's making us such a dysfunctional group.

> I've never seen such a ruthless development team.

It's the sort of thing that makes me wonder what I'm doing in the profession.

At first it seems like an attentive gesture—he is responding to my every line—but soon I feel as though I am living with an echo. Not only do I get a response back in a hurry, but what I get back are *my own words*. I would rather see what he remembered of my mail. I would like to know the flow of his mind, how it leaps from one paragraph to the next. But instead I get interpolations. I don't feel answered; I feel commented upon. I get irritated, should say something, as one should in any relationship. But I let it go, just break a thread (I don't type "r," dropping his subject on the "Re:" line) to signal my displeasure.

Months go by. Slowly, without ever talking about it, we work out the interpolation problem. We get good at it, use it. I write to thank him for recommending a book, and he interpolates his reply:

Thanks again for the book. I don't want to finish it.

My pleasure.

I like having it by my bedside.

My pleasure.

—E

—K

Meanwhile, our daylight life moves in a separate, parallel track. When we "speak" in the group alias, it's without overtones. I even report a bug in Karl's code as I would in anyone's. When I have to write to him directly about some work matter, I always "CC" the lead engineer. The "CC" is the signal: Watch out, pretend you know nothing.

Only once does our private world intersect with our work. I have to get a technical particular from Karl; mail would be too slow, I use the phone. I say my name, and our voices drop to a soft low tone. I am talking about a program—"So it becomes 'root' then calls 'setuid' to get read/write/execute permissions on the file"—but I am murmuring. In my mouth, "root" and "call" and "permissions" become honeyed words. He responds slowly. "Yes. That's what it does." Pause. Low talk: "All permissions. Yes."

Exquisite as delay has been, we can't put off meeting indefinitely. The email subject heading for the past month has been "Dinner?" and we both know we can't keep writing messages under this topic and never actually have dinner. Perhaps it's simply the way words have power over our software-engineered lives: The dinner date sits there as a mail header, and we have no choice but to fulfill it. There are real and good reasons we should resist the header, why we should stay where we are and just send mail. We work together. We're both just out of long-term relationships (which we've discussed). I've tended to prefer women (which he doesn't know; I'm not even sure what I should be telling myself about all this). Still, there is a momentum by now, a critical mass of declared "we-ness" that is hurtling us towards each other. It must be done: We will have dinner.

By the time he is to arrive, my body is nearly numb. Part by body part turns off as the time for his actual presence comes nearer. He calls. He's going to be late—bug in a program, traffic. I hear the fear in his voice. It's the same fear as mine: We will have to speak. We will have to know when to talk and when to listen. Panic. We have no practice in this. All we know is we must type "r" and reply, reply right now. Without the press of the system, how will we find the auditory, physical rhythm of speech?

We should not have worried. We sit down in the restaurant, and our "conversation" has an all too familiar feel. One talks, stops; then the other replies, stops. An hour later, we are still in this rhythm. With a shock, I realize that we have finally gone out to dinner only to *exchange email*. I can almost see the subject headings flying back and forth. I can even see the interpolations: "About what you said about...." His face is the one of my imaginings, the same serious attention, deep voice, earnest manner with an occasional smile or tease. But, in some odd way, it's as if his face is not there at all, it has so little effect on the flow of "talk." I look at our hands lying near each other's on the table: They might as well be typing.

We close the restaurant—they have to vacuum around us. It's nearly midnight on a Tuesday, and he gives off the cues of a man who has no interest in going home. He says "Yes, the beach" before I can even get to the alternatives of the Marina, the new pier at the Embarcadero, a South of Market club. Yes, the beach.

A storm is coming in off the Pacific. The air is almost palpable, about to burst with rain. The wind has whipped up the ocean, and breakers are glowing far out from the beach. The world is conspiring around us. All things physical insist we pay attention. The steady rush of the ocean. The damp sand, the tide coming in to make us scuttle up from the advancing edge. The sandpipers busy at the uncovered sand. The smell of salt, of air that has traveled across the water all the way from Japan. The feel of continent's end, a gritty beach at the far edge of a western city.

Yet we talk, talk, talk. My turn, over; your turn. He walks, briskly, never adjusting his pace to mine, and he talks, talks, talks. Finally, I can't stand it. I just stop. I put my hands in my pockets, face the ocean, and watch the waves setting up in the dark. I feel my whole body saying, "Touch me. Put your arm around me. Only brush my shoulder. Even just stand next to me, your hands in your pockets, but our jacket sleeves grazing each other."

Still we march up and down the beach. He clearly doesn't want to leave. He wants to stay, talk, walk at that relentless, never-adjusting pace. Which should I believe: his staying with me at midnight on a deserted stormy beach or this body-absent talk?

Across the road from the beach is an old windmill that doesn't turn, a *folie* of the 1890s. Naturally he is interested, he wants to go there, walk, see everything. I tell him I think it once worked, something about an acquifer under the park and the windmill used to pump up water. We think it over. It's consoling, this engineer talk, this artifact of a thing that once did actually useful labor, handiwork of the Progressive Era, great age of engineering.

Surrounding the windmill are tulips, white, and a bench. I want to sit quietly on the bench, let my eyes adjust to the dark until the tulips glow like breakers. I imagine us sitting there, silent, in the lee of a windmill that doesn't turn in the wind.

But I look up to the top of the windmill, and I can't help myself.

"A dish!" I exclaim. What appears to be a small satellite dish is perched in the spokes of the mill.

He looks up. "Signal repeater," he says.

"Not a dish?"

"No, signal repeater."

It is kind of small for a dish. He's probably right. "I wonder what signal it's repeating," I say.

We're finally quiet for a moment. We look up and wonder over the signal being repeated from somewhere to somewhere across the ocean.

"Navigation aid?" I hazard. "Marine weather?"

"Depends," he says. "You know, signal strength, receiving station location."

I think: antennas, receiving stations. Spectre of hardware. World of Eugenes. Bits and protocols on air and wire. Machines humming alone all night in the dark. "Yeah," I say, remembering the feel of CQ through electric snow, giving up on the evening, "signal strength."

Near dawn, I'm awakened by the sound of drenching rain. The storm has come in. My cat is cold, scratches at the top sheet for me to let her in. We fall back to sleep like litter mates.

For a few hours the next morning, I let myself feel the disappointment. Then, before noon, the email resumes.

He writes. His subject heading is "Thank you!" He thanks me for the "lovely, wonderful" evening. He says he read the article I gave him before going to bed. He wanted to call me in the morning but didn't get to sleep until 2 a.m. He woke up late, he says, rushed from meeting to meeting. I write back to thank him. I say that, when we walked on the beach, I could smell and feel the storm heading for us across the Pacific. How, when the rain's ruckus awakened me in the night, I didn't mind; how I fell back to sleep thinking to the rain, I was expecting you.

Immediately, the body in the machine has returned us to each other. In this interchange there is the memory of the beach, its feel and smell, mentions of beds and sleep. Bed, a word we would never say in actual presence, a kind of touch by word we can only do with our machines. We're programmers. We send mail. It's no use trying to be other than we are. Maybe the facts of our "real" lives—his ex-girlfriend, my ex-girlfriend, all the years before we met in the group alias— mean we won't touch on deserted shorelines or across dinner tables. If so, our public selves will go on talking programs and file permissions in a separate and parallel track. If so, we're lucky for the email. It gives us a channel to each other, at least, an odd intimacy, but intimacy nonetheless.

He ends with, "We should do it again soon..." I reply, "Would love to." *Love to.* Who knows. The world is full of storms and beaches, yes? Below, I leave the two interpolated signatures:

—K

   —E

The Associated Press reports that the Coast Guard has turned off its Morse code equipment.[5] At 7:19 p.m. on Friday, March 31, 1995, stations in Norfolk, Boston, Miami, New Orleans, San Francisco, Honolulu and Kodiak, Alaska, made their final transmissions and simultaneously signed off. "Radiomen" would henceforth be called "telecommunications technicians." The dots and dashes of S-O-S would no longer be the universal message of disaster. Ships at sea would now hear about storms and relay distress signals via the Global Maritime Distress and Safety System, which includes "a satellite-relayed signal giving the ship's location"—many signal repeaters lining America's beachfronts, no doubt.

Veteran radiomen gathered to mourn the passing of the Morse code. "Dots and dashes are probably the easiest things to detect bouncing off the atmosphere," said one; and I remembered how, on stormy nights, Eugene would resort to code, which he liked to say aloud as he transmitted, pronouncing it "dit-dit-dot, dit-dot." One ten-year radioman, Petty Officer Tony Turner, talked about losing the feel of the sender. The transmission comes "through the air, into another man's ear," he said. The code has a personality to it, a signature in the touch and rhythm on the key. For Turner, the signature's origin is no mystery. "It's coming from a person's *hand*," he said.

Endnotes

. . . .

1. FTP stands for "file transfer protocol," a standard method for transmitting computer files between machines on the Internet. The "IP address" is the Internet Protocol address, the unique identifier of an Internet-connected computer, or "node." After transferring a file from an FTP site, the recipient usually creates local files and directories using a program called "tar," which creates archives. (The program name is derived from "tape archive"; tapes are rarely used these days, but the device lives on in the command name.) "Tar files" are therefore named after the program that processes them.

2. The names of email commands themselves demonstrate why and how UNIX has become "arcane." A logical, clear name for an email command would be something like "electronic_mail." However, UNIX is designed for programmers, people who type, and the goal is to create program names and commands that can be invoked with the fewest number of keystrokes. Therefore, "electronic mail" became "elm." The problem is that the shortened name acquires a new association: trees. So, when a later email program came along, it was called not "mail" or "email" but, inevitably, "pine." Longtime users don't find it at all strange to use a program named "pine." But novices certainly think it's odd to start sending mail with the name of a tree.

3. There are many "visual programming" tools now available which enable users to create programs with a minimum of coding in computer languages. In place of code, programs are assembled mainly by clicking on icons and dragging things around on the screen. However, these visually oriented tools are not in the workbench of software engineering; engineers may create such tools but generally do not use them. Increasingly, the engineering language of choice is C++, which even longtime engineers find syntactically "ugly." Below is a C++ code sample. It is excerpted from a programmer joke called "The Evolution of a Programmer," which has been circulating on the Internet. Although part of the joke is to make the code as complex as possible (it is possible to write these instructions more simply and clearly), the sample does demonstrate how far from point-and-click lies the language of software engineering.

```
#include <iostream.h>
#include <string.h>

class string
{
private:
   int  size;
   char *ptr;
```

```
public:
   string() : size(0), ptr(new char('\0')) {}

   string(const string &s) : size(s.size)
   {
      ptr = new char[size + 1];
      strcpy(ptr, s.ptr);
   }

   ~string()
   {
      delete [] ptr;
   }
   friend ostream &operator <<(ostream &, const string &);
   string &operator=(const char *);
};

ostream &operator<<(ostream &stream, const string &s)
{
   return(stream << s.ptr);
}

string &string::operator=(const char *chrs)
{
   if (this != &chrs)
   {
      delete [] ptr;
      size = strlen(chrs);
      ptr = new char[size + 1];
      strcpy(ptr, chrs);
   }
   return(*this);
}

int main()
{
   string str;

   str = "HELLO WORLD";
   cout << str << endl;

   return(0);
}
```

The code above prints the words "HELLO WORLD" on the screen in plain text. The very same functionality can be accomplished using BASIC (a programming language no longer used by "real" engineers) in a program containing the following two lines:

10 PRINT "HELLO WORLD"

4. The current paradigm in software engineering is "object-oriented programming." In this model, programs are designed and written in units (encapsulated segments of instructions and data) that can be related and reused in complex ways. Objects are combined to create components at various levels of granularity, from a small object that checks a single character to one that runs an entire spreadsheet routine. Although there is much obfuscation in discussions about object orientation, it can logically be understood as an attempt to treat software as if it were hardware—as assemblies of standard parts.

5. Joe Taylor, "End of Morse," 31 March 1995, The Associated Press. Dateline Norfolk, Va. Emphasis added.

The Memoirs of a Token: An Aging Berkeley Feminist Examines *Wired*

Paulina Borsook

I suppose it all came to a head one night past midnight in the spring of 1995. I lay awake on the couch in the living room thinking dark thoughts until past three in the morning. *Wired* magazine had brought me to that too-often described "aha" state of women and ethnic minorities and gays and anyone else who feels marginalized and badly done by, when they reach the epiphany that requires leaving policies of appeasement behind—wounded pride, self-disgust, the infant impulse to lash out.

Wired is the hottest, coolest, trendiest new magazine of the 1990s. Within its first three years, it has received a National Magazine award, started sister publications in England and Japan and become the coffee table ornament/ lifestyle indicator/print-based fashion accessory *du jour.* This consumer magazine that makes computers and communications and video and technology sexy and glamorous even shows up in IBM's television commercials. *Wired* marks its readers as being with-it one-planet high-tech high-touch global citizens of the 1990s.

And *Wired* has certainly been the most successful magazine launch San Francisco has seen since *Rolling Stone* in the mid-1960s. Like *Rolling Stone*, it seeks to both shape and reflect an emerging, media-driven worldwide culture, one that's creative and anarchic and playful and loose in the hips. And like *Rolling Stone*'s founder, Jann Wenner, *Wired* co-founder Louis Rossetto is well on his way to creating a media empire of his own, everything from the online *HotWired* to HardWired Books to rumors of Wired TV.

Memories of Underdevelopment

But here's where things get tricky and uncomfortably gendered: When *Rolling Stone*, the chronicle of sixties sex'n'drugs'n'rock'n'roll culture came into being, sixties feminism hadn't really yet hit the streets. Hippie chicks were all into free love and making bread, and Janis and Grace sure could wail; but somehow girls weren't much relevant (o, sixties concept!) to making it all happen. The second wave of feminism hadn't yet spawned its kazillions of cultural commentators (How many feminists does it take to screw in a light bulb? One to do it, and four to write about it).

The counterculture was supposed to affect us all—however, you wouldn't know that women were up to much, really, after reading *Rolling Stone*. Women getting uppity (as they did as the sixties turned into the seventies and into the eighties) were intrinsically *not* considered entertaining—nor of much interest to the fifteen-year-old boys (*Rolling Stone*'s true subscriber base) who read the magazine in the rec room of their parents' basement, and to whom the magazine represented the fantasy lives they wished to live: hip, smart, irreverent. *Rolling Stone*, like *Wired*, like all consumer magazines, acts out its readers' fantasy lives (consider that half of the subscribers to *Vogue* are size 12 or larger, and that many of those who subscribe to poetry magazines are would-be poets who hope to get published themselves someday).

Thirty years have passed since Jann Wenner started *Rolling Stone*, yet *Wired* , its spiritual child, both interpreter and vanguard of the Next New Generational Thing, is also turning out to be largely by guys and for guys. It's as if women hadn't infiltrated, at least to some degree, every kind of traditionally male bastion since the 1960s—and as if it wouldn't be perilous and foolish and shortsighted to conceive of and record any kind of global transfiguration of culture that didn't include them.

We're not quite as invisible or handmaidenly this time around, but *Wired* doesn't seem to have paid much mind. We have in us, if only as ancestral genetic memory, the experience of women civil-rights workers who began to wonder why they weren't allowed to participate in the revolution they came to foment—and who came to demand a commensurate place in it, and in their own lives overall.

Nonetheless, *Wired* has consistently and accurately been compared in the national media to *Playboy*. It contains the same glossy pictures of certified nerd-suave things to buy—which, since it's the nineties, includes cool hand-held scanners as well as audio equipment and cars—and idolatrous profiles of (generally) male moguls and muckymucks whose hagiography is not that different from what might have appeared in *Fortune*. It is the wishbook of material desire for young men.

And as with *Playboy*, whose articles can be both *great* and about things that count society-wide, women tend to be put off by *Wired*, as they seem not to be by *Wired*'s more wacked-out sister-spirit competitor publications, *boing boing* and *Mondo 2000*. When thirty women at a San Francisco talk about women and the Internet (a pretty self-selected bunch relatively comfortable with technology) were asked if they read *Wired*, they all raised their hands. And when they were asked if they hated *Wired*, they all raised their hands. It appears women, even the techno-initiated, generally do *not* like *Wired*.

Looks Count

Which poses an uncomfortable question: If something looks sleek and whiz-bang, does that mean it *can't* appeal to most women? That is, if a publication exudes the smell of new machines, celebrates the whizzy and the zippy—are there subtle gendered cues that say "Boys club! Fun for us! None of that cuddly touchy-feely serious crap that reminds us of moms/girlfriends/schoolmarms!" And does that also telegraph to women the message "trancelike boy-absorption in silly fetishizing junk on the order of model airplanes—who needs it?"

Would art directors and style-setters worldwide ape the *Wired* style if it telegraphed *girls* just having fun? Put the *Wired* visual gestalt in the same box as other mostly-male-reader mags such as *Popular Science* or *PC Magazine.* If they looked like *Ladies Home Journal* or *Martha Stewart Living,* would male readers refuse to pick them up? Why aren't the similarly tacit appeals to out-there lifestyle choices of *Interview* as gendered as those of *Wired?* Is it because *Interview,* with just as tendentious an à la mode look and feel as *Wired,* is nominally about *people* and *Wired* is about *machines?*

The Second Sex

The very success of *Wired* raises another unpleasant revenant: It's a sociocultural commonplace that when a profession, hitherto a male dominion, becomes more accessible to women, it loses status. Clerks, bank tellers, secretaries, pharmacists, schoolteachers, English professors, psychiatrists, were at one time practically all men—these are occupations that lost money, status and power as they became more female-dominated. Does this mean that a trade and a subculture can't possibly be of the moment and desirable and high-caste and way-cool and possessed of societal *juju* if it is seen as being female? Or even, female-friendly? It's a truism that women wear male clothing (either the real thing, or reinterpreted for them), but few men wear women's clothing—unless they are into gender-fuck or transgressive behaviors in general. A certain sexually charged salability aside (Swedish bikini team or dyke chic), girl-associated culture isn't perceived as cutting edge.

Wired says that 20 percent of its readership is women. Would *Wired* ever run an article of more appeal to women than men, just to favor for once one of those 20-percenters? Would its core readers feel betrayed?

Material Culture

The wild success of *Wired* also touches on other equally

creepy gender-oriented uneasinesses. A peculiarly awful, misogynistic boss I once had at a now-defunct computer magazine in the early 1980s once made the dismissive remark that no general-interest magazine whose audience was mostly women could survive commercially (witness the problems with *Ms* and *Psychology Today*)—unless it was a female service-magazine such as *Redbook*. But magazines whose readership was largely male (such as *PC World*) could flourish. That ignores all the niche hobbyist magazines largely for women, and the nominally-unisex-but-really-for-women titles such as *Vanity Fair* and *Gourmet* and elides the important distinction that computer magazines, in some sense, are very much business and trade publications.

But his point still nags: Since women generally don't fetishize and aren't tool and object junkies to the extent that men are (though of course there are exceptions), maybe there can't be a magazine that guarantees to a set of advertisers (and it is advertisers, and not subscribers, who subsidize a magazine) a captive audience of readers interested in particular kinds of goods and services they have to sell. The idea is, women aren't interested in objects, unless it's the bland health/grooming/fitness/beauty/Kitchen Bouquet chachkas that show up mostly in the pages of rags such as *Good Housekeeping*.

You Do the Math

When all else fails, you can, as they say in Internet slang, RTFM (read the fucking manual). That is, IRL (in real life), examine the text at hand. I sat down with all *Wired* issues published as of late spring 1995 and performed the lame, plodding, numerical analysis so annoyingly a part of feminist studies and employment-discrimination class-action lawsuits. I counted how many men and how many women were authors of *Wired* stories and subjects of *Wired* features, and how many were listed on the *Wired* mastheads.

It would be intellectually dishonest to expect *Wired* to practice what a male *Wired* editor calls "affirmative-action journalism," that is, to expect more women to appear in the magazine than actually scrabble around in the world of digital convergence. The best figures available seem to indicate that about one third of Internet accounts are now owned by women, but historically that has not been the case. And it's clear to anyone who spends time online that even now women aren't taking up one-third of the bandwidth. And only about 10 to 15 percent of computer-science majors (both graduate and undergraduate) are women.

And thus it appears: Women were the subject of about 15 percent of *Wired* articles and only one cover out of twenty-five as of June 1995. There are vague in-house grumbling intimations by women on staff of likely accomplished females being shot down as article-subjects in *Wired* editorial meetings and of no-account off-point numb-nut guys being celebrated instead—but that's par for the course in any mostly-male arena.

Where *Wired* numbers become dismaying lies in who is doing the article writing. The community of writers capable of writing for *Wired* is not made up of the same kinds of people who might be considered net geeks or multimedia content developers—or who in their fantasy-life might want to imagine themselves to be such.

Schools of journalism overall attract and graduate women in equal or greater numbers than men; even the staffs of male-oriented magazines such as *Playboy* are made up of sizable numbers of women. What's more, within the specialty of technical journalism, women make up 30 to 50 percent of its work force, much as women dominate technical writing and documentation—because that's where the jobs are.

But only 15 percent of *Wired* authors are women.

Because computer journalism as a specialty has been around for at least ten to fifteen years, plenty of women have been able to acquire the chops to make it to the journeyman, and even expert, ranks of this specialized writing guild. But these women seldom show up in the pages of *Wired*.

These dreary *Wired* statistics fit with what many, many women have told me: that they had pitched *Wired* and gotten nowhere, not even the courtesy of a formal rejection.

Is it that women who habitually *do* write about technology simply can't phrase their queries in ways that *Wired* guys like? Is it their choice of subjects? Is it that they look at the magazine, think "boys and their toys; no way they'd want *me*" and don't even try? And as for the women who are simply damned fine writers and could write about anything if asked (you know, like Janet Malcolm or Katha Pollitt or

Cynthia Heimel)—are they not asked? Would they laugh if they were? Would the whole magazine strike them as being inaccessible and adolescent-male-wet-dream and, therefore, not worthy of their efforts? Highly paid, highly visible male writers want to be in *Wired* regardless of the pay because it's the current guy magazine of record, whereas women writers of similar status appear to not know it exists, not care, or maybe figure it's too much of a boy's club to even bother.

Historically the editorial positions of creativity, power and prestige at *Wired* have been held by men, while editorial pooper-scooper jobs, the copy editors and fact-checkers and everyone else who has to clean up and rewrite copy into acceptable shape for publication, have been held largely by women. They have contended with substandard prose written by male writers all the time (where'd they get these guys?). Are the women who can't even get in the door worse than the guys who are already in? Do the women's shoes have to be shined twice as bright? four times as bright? eight times? to compete with the boys whose output makes *Wired* staffers' eyes roll?

As for the women authors who do make it into *Wired*, they seem to write an inordinate number of articles on sex'n'dating, or profiles of other women; can they not think of anything else to write about? Or is this the equivalent of what happens in big corporations, where no one knows if it's due to women unconsciously seeking their comfort zones or men unconsciously ghettoizing their female hires? Women in management inevitably show up in human resources ("so good with people"). In the legal community they end up in family law, in urban planning they cluster in housing.

As for the masthead, it reflects who a magazine considers its stable of contributors to be, even if they haven't written a thing in months: It's a way to list membership in its club of the cool, the equivalent to "of counsel" on a law firm's letterhead.

Sad to say, in my survey of *Wired*, women made up only 8 percent of the writers listed on the masthead. There was one month when I was the only woman on the masthead. Month to month, I have been in the company of only one or two others.

Does this issue of the gender of authorship matter? To me gender and agency are inextricably bound up with each other, though necessarily not in a one-to-one causal relationship: For example, it's still the case that women legislators, regardless of party affiliation, care more about childcare and education and family-leave than male legislators do. And just as in fiction, where men trying to write from a woman's point of view can create fascinating work that to me almost always rings false, I think it does matter that so few of the voices printed in *Wired* are female. It probably doesn't occur to the men writing for *Wired* to think about issues that might be of more importance to women than to men.

The Personal is the Political

Then there's my own history with *Wired*, which explains how I ended up such a wreck that night on the couch. Of course, all kinds of folks feel marginalized and dismissed by the *Wired* sensibility and the *Wired* powers-that-be. Indeed, my sense of the magazine's rejecting me with a foreign-body reaction was probably just as much a function of my own intolerance of their wacky, techno-naive libertarian worldview (technology = good, government = bad; free market = good, regulation = bad; digital = good, analog = bad) as it was the magazine's trivialization of the token girl. Although *Wired* maintains a posture of celebrating all the cacophony and lack of prior restraint net global culture has to offer, in fact, it is not open to points of view other than its own, as bounded a set as *National Review* or *The Advocate*.

Ultimately, gender and gender politics are simply part of the political mix of all humans operating in groups, and not everything can be squished through the narrow cognitive-filter of feminist critique.

That being said, commence my lament:

Back in the winter of 1993, before I'd returned to live in San Francisco, I had dinner with John Markoff, computer correspondent for the *New York Times*. Markoff is someone I've known more socially than professionally, if tangentially, for years, and he asked me if I knew about these nice, interesting people at this new startup, *Wired*, which had billboards on buses all over town. No, I hadn't heard of them, but Markoff had written for the magazine's first issue, so I figured the founders might be engaged in something decent.

A few weeks later, I received a free copy of the first issue of *Wired* through a one-time mass mailing. The magazine had obtained my name and address from the membership roster of the Electronic Frontier Foundation (EFF), a non-profit cofounded by Lotus 1-2-3 entrepreneur Mitch Kapor and Grateful Dead lyricist John Perry Barlow and dedicated to the preservation of First and Fourth Amendment rights in cyberspace; I'd been a charter card-carrying member.

I describe this periphrastic backstory so that it's clear *why* I thought *Wired* was an association of kindred spirits, communicators whose larger view on technology was not so different from my own. At its start, and at the commencement of my connection with it, *Wired* and I seemed to be cohabiting the same small sector of the galaxy, and there seemed a certain synchronistic rightness to the connection. But I digress...

When I looked over that first issue of *Wired*, with its glossy literate tone (plus a Q+A with Camille Paglia, the counterfeminist of the moment), I thought, "Hmm, maybe they'd get my fiction." My MFA thesis written for Columbia University's School of the Arts was a series of interconnected short stories depicting how the new information technologies deform relationships, focusing on but not limited to email. I'd been receiving Miss Congeniality award rejections on it for months, often written on manual typewriters ("you're a very good writer, and this is very interesting, but no one cares about this email nonsense and this is all unrealistic anyway").

After asking editor in chief Rossetto by email if he'd be interested (in those days, he was easily reachable), I asked my agent to FedEx the manuscript. They received it on a Friday and agreed to publish its concluding novella, "Love Over the Wires," the following Monday. I was told they were thrilled to have a Pushcart Prize-nominated writer publishing fiction with them, and they started throwing nonfiction projects my way.

Working on my first *Wired* profile, a portrait of the Hannah Arendt/Jeane Kirkpatrick of the computer industry, Esther Dyson, I included a section on gender-politics, because I had been struck in my reporting by how often I was asked "Who is Esther dating?"—a question that had *never* come up when I had done profiles of other computer folks, that is, guys. I was also struck by the personal quality of what her detractors had to say; that is, people made snotty remarks about her clothes or manners that I had once again never heard applied to equally socially aberrant guys. And finally, Esther was interesting as a woman who had succeeded in a male-dominated field.

Making Esther neither a feminist exemplar nor a martyr, I simply described in a limited portion of my article the comments and questions about her that I had found curious, indicated that such an anomalous set of reactions had to be related to her gender, and asked Esther how she felt about the ways her life-path had been affected by her being a woman.

Executive editor Kevin Kelly called me up to complain. Wasn't it sexist of me to call attention to Esther's gender at all? "I don't even think of whether or not Esther is a woman." Of course he didn't; Esther-the-uncoupled-off-tomboy is one of the best women I've ever met at being one of the guys. Wasn't it irrelevant of me to bring up all these feminist issues? I responded that it would be of interest to other women to see how a woman negotiated all the compromises and contradictions of making it in a male-majority polis. "Gee, I never thought about how it might be interesting to women to hear about another woman's experience." And on it went, back and forth. That section remained; it is the part of the article I have gotten the most compliments and comments on; but, of course, as Deborah Tannen has pointed out in her books, anytime a woman raises an issue of gender-politics, or a relationship issue, she is branded as troublesome, regardless of how polite or righteous her cause.

I'd never had a conversation quite like this with an editor before; puzzled, I trotted off to...

The Salon

I was asked to participate in a roundtable discussion with various people considered Interesting and Worthy by *Wired*, perhaps modeled on the monthly Reality Club meetings of New York literary agent John Brockman (who handles clients such as *Whole Earth Catalog*'s Stewart Brand and John Markoff). I was amused and flattered and thought it would be fun.

Alas, what resulted was one of those click-inducing,

radicalizing smashups I'd heard about and read about since 1969 but never before experienced for myself.

The Friday-night salon turned out to be just like those conclaves that women who attend graduate school in male-dominated fields in science report on and that women in the business world complain about. In other words, no one paid me, one of the few women invited, the smallest amount of attention.

Some of my problems that night were no doubt structural: My female tendency towards rhetorical politeness (not wanting to barge in or raise my voice to overmaster someone else's speech) and my softer voice, were, I'm sure, part of the problem. And a male friend I ran into later commented on the bombastic, bad energy in the place that night, so gender probably wasn't the only determinant of bottom-dog status.

Nonetheless, at this confab at *Wired*'s South of Market offices, I was rendered mute. When I would start to talk, the other salon attendees would simply act as if I weren't present. If I made a point, it was simply ignored, evidently not worth responding to. One participant was a guy I had coincidentally interviewed on the phone a few hours before for a non-*Wired* project—he acted as if I were not even there.

The climax of the evening came when virtual-reality spokesmodel Jaron Lanier started babbling on about how "we" (who we? *Wired?* the technology community? self-absorbed guys with large and fragile egos? monomaniacs?) needed to seek out artists and humanities people, people whose comments and critiques were not normally heard. Meanwhile I had been sitting right next to him all evening, and he hadn't even acknowledged the volume of air I was displacing. I felt like grabbing him by the dreds and saying, "Asshole, I studied with Kenneth Rexroth and published poetry in little magazines whose names even I have forgotten! I've been hanging out with arty types like Leo Castelli hangers-on at places like Max's Kansas City since 1973!"

But I couldn't see what such defensiveness and such protestations and such borching would accomplish. I was appalled at having been put in a position where it was assumed I was a null set, useless until I proved otherwise. Not being inclined toward scenes, I said nothing at the time. I did throw a hissy fit afterwards (for I had never been ignored just like that), and naturally, was not invited back.

The events at the *Wired* salon were a sign that something very odd was going on with this community/culture/cult. I hadn't expected bleeding-edge guys from their twenties to their fifties to act like the board members of General Motors sometime in mid-1957, and it's certainly not been borne out by my encounters with the Real World Out There that technophiles are, to a man, male chauvinist pigs. The manners of the *Wired* salon participants were quite different, say, from what I had run into with the friendly geeks of the Internet technical community. Those Internet technocrats were attentive, verging on courtly, once they saw I meant no harm and was at least educable.

Tending toward the dense, I couldn't reconcile the cognitive dissonance between *Wired*'s representation of itself (egalitarian, democratic) and *Wired*'s operational nature.

Cut a year later to...

Femxpri

In the summer of 1994, a woman started the topic "WIRED...gendered? (oh no!)" in the *Wired* conference on the WELL, the online conferencing system/BBS I've lurked on for years. The predictably accurate but hopeless comments were posted: the aggressive, swaggering look and feel of the mag (most likely appealing to guys, probably off-putting to gals), the emphasis on expensive toys likely to induce male vasocongestion and capable of being used as display and territorial markers (women, to the extent they care about such things at all, generally are interested in the utilitarian value, not fetish-worthiness, of tools), the generally unconscious but annoyingly present small sexisms, and so on.

Publicly I posted a few remarks, mostly to the effect that guys have this amazing "off" switch that makes them simply not want to hear women complaining about men, or talking about topics exclusively of interest to women—unless it's about sex in a way guys find exciting and not threatening. But what signified more is that while I was posting that publicly, I wrote what I *really* felt in a private women's conference on the WELL, femxpri (a conference at least nominally for Generation X women). This public-private duality in part sprang from my not wanting to violate the confidences told to me by female *Wired* staffers. I also didn't want to jeopardize my position at the magazine. Not totally. Not at that time. *Wired,* like many entrepreneurial organizations, is very much of the either-you-are-with-us-or-you-are-against-us mentality. It pronounces anathema on those who defect for any reason and cannot abide criticism or deviation from its tiny cel of the world-kaleidoscope. I knew to write all I thought and felt would end my career as The Woman Writer They Respect and Can Point to When People Accuse Them of Being Sexist. I was well aware, like many tokens in high-profile scenes before me, that I had a position of some status that I enjoyed and wasn't ready to relinquish.

This dodge is somewhat allied to the phenomena I'd observed in social or professional situations where it came out that I'd written for *Wired.* Suddenly I would be perceived to be more interesting/intelligent/attractive than I had been in the seconds before this revelation was made. But sickeningly enough, I came to feel this highly favorable response may have had *less* to do with the reflected sheen of *Wired* and more to do with the oddity of such an unprepossessing creature (a frowzy, lumpy, bookish, middle-aged woman) creating content capable of appearing in *that* mag. Didn't Dr. Johnson make a misogynistic remark to that effect: that women preaching is on a par with a dog's walking on its hind legs; the marvel is not that they do it well, but that they do it at all....

Which feeds, sheepishly enough, into a confession regarding...

Writer Vanity and Thin-Skinnedness

Anyone who works for applause (actors and film editors and writers) cares if his or her work is displayed to maximum advantage, is leveraged for optimum narcissistic gratification. Which in the negative, means it's really easy to start throwing all manner of slights into one's own private cauldron of discontent: Never a cover story. Screwups on layout and design and type. Not being invited to A-list parties. Unreflective slapdash press-release journalism rewarded lavishly in others. Never being paid on acceptance, as my contract stipulates.

As a consequence, it's hard to disentangle the inevitable

gripes stemming from the general *Wired* cavalier attitude toward its writers (I know of no one, regardless of gender, currently writing for them that is happy with the magazine); the personal-to-me lack of fit between my worldview and that of *Wired;* and the instinctive antipathy I've always had toward star fucking and status seeking, which *Wired* is tending toward more and more. But considering I've been with the damned thing from the beginning, that I've been told repeatedly that I am valued as one of their best writers (when I was applying for a job at *The Economist* in the spring of 1994, the managing editor of *Wired* told the Brits that they could have me only if I was still allowed to write for him), why did I feel shunted aside? I'm not that paranoid....

Which leads to the blowout/couch-huddling/coup de grâce incident triggered by women technopagans.

Around the same time as the femxpri discussions on the WELL, I finished writing a controversial profile for *Wired*, this one of Paul Allen, the other billionaire founder of Microsoft. This was a project beset by false starts, blind alleys, confidential sources, massive use of the computer industry's bush telegraph, and reliance on the sources—and the reputation for accuracy and fairness—I'd built up over years. It was investigative reporting of the classic kind: as such, it got lots of people upset and required the services of the *Wired* libel lawyer—and exposed me to serious censorship. Rossetto deleted material that didn't fit his ideological praxis. The article took six months, eight drafts, and was exhausting for all parties concerned.

So for my next project, I wanted to do something fun, where no one would be disposed to get mad and no moneyed interests were at stake. It was a story idea I'd had in mind for a long time: technopagans, high-tech folks practicing neopaganism, nerds making like wizards on their nights off or actually using technology in their rituals. I thought they were a hoot, and I relished writing as an inquiring anthropologist from Mars about a subculture I found entertaining. And it seemed like a pure-form emergent-culture *Wired* feature.

Wired hemmed and hawed; wouldn't say yes, wouldn't say no. I kept presenting my case to John Battelle, the managing editor, who had always served as my first-line editor. He was lukewarm at best.

But at the same time, in a fit of pique + divine madness, my then-roommate Andy Reinhardt (fellow humanities-geek/ executive editor for *PC World*) and I wrote a spoof treatment of a TV show, "beverly_hills.com." It was silly and satirical and in the tradition of a *New Yorker* literary parody, sketching out the madcap doings of a gang of cool Gen X kids as they searched the Information Superhighway for love and adventure.

Still undecided about technopagans, Battelle dithered about actually reading "beverly_hills.com" for weeks. When he finally did read it, he quickly snagged it for the magazine. But he had also decided to give the technopagans assignment to a *Village Voice* writer, Erik Davis, whose query came in after mine. Doing a little damage control, Battelle suggested we go out for dinner and a chat.

At dinner, he told me that *Wired* had been blown away by "beverly_hills.com"; hadn't realized I was capable of playfulness (in other words I had been typed as the conscientious but humorless shrew; remember what Deborah Tannen said about the bad rep that attaches itself to women who kick up a fuss?), which presented the magazine with a quandary: I *was* on the masthead and *did* have seniority and *had* pitched the technopagan idea first; nevertheless, they had wanted to give the very-cool Mr. Davis a chance in the magazine.

Implicit, of course, is that my pitching the proposal didn't make it a worthwhile subject. As an aside, it turns out that another writer, another woman, had tried to pitch the same story to *Wired* months earlier than I had and had gotten even more nowhere than I had. But when Davis pitched the story, it suddenly became more legit. But "beverly_hills.com" for *Wired* revealed another side of my writing, and now *Wired* was chagrined. I found it puzzling (and to be honest, I was miffed) that they hadn't thought me capable of lightness or deftness or insouciance: I had been communicating with them, by phone, email and in person for two years; they'd been reading and editing my writing for two years; they'd read my fiction—sheesh!

Battelle suggested I do a sidebar to Davis's story. Davis okayed this and came up with the idea of a sidebar on women technopagans. Davis has impeccable feminist-male credentials and is as tuned into gender politics as a guy could be; and he and I were equally cognizant of, and uncomfortable with, the notion of ghettoizing the female aspects of the subject into their own pink-beribboned little add-on. We agreed that, best case, women would have been woven into the story naturally (unusual in tech culture, women are key players in technopaganism), but that assigning me the femme concerns was a natural and easy way to create a companion piece to round out the narrative.

Anyway, as awkward as it was for me to sequester the women pagans off by themselves (the stealth affirmative-action journalism I practice means making sure I include as many women sources as I can in a story, so that women are heard, but weaving female issues and ideas and experiences in where it fits—the best way to combat sexism is to render women visible, but not to make a big deal about them as Other), I was happy to have *part* of my story back and glad to sneak some women-centric writing into the mag.

Battelle promised that Davis's piece would run around three thousand to four thousand words, and mine around fifteen hundred words.

So I wrote my piece on girl technopagans, careful as I always am not to engage in childish "woman = good, man = bad" false dichotomies (if anything, as a straight but not narrow het who from one year to the next is likely to have more male chums than female ones, I am a male apologist and insist on seeing and describing the world in all its non-binary shades-of-gray nuance); Davis wrote his main bar; he and I talked and swapped proto-drafts. His piece had to undergo a massive rewrite, so I heard nothing about my "The Goddess in Every Woman's Machine" until months after I turned it in.

At the last minute, arriving by email after 11 p.m. one night, only a few days before the package of articles was to head off to its final staging areas of layout and art and fact-checking and the printer—and I was off to New York for a business trip, and consequently had little time to do much squawking—I received "The Goddess in Every Woman's Machine" from managing editor Battelle. It has been cut to five hundred words and was annotated with the comment that he wasn't sure he could salvage it at all with executive editor Kelly and editor in chief Rossetto, especially since it was so different from what Davis had written.

I went berserk. I *knew* that if the article had been on

cryptography and technopagans, hot sex chat and techno-pagans, any *Wired* hot-button topic and technopagans—hell, pretty much anything other than women technopagans—there wouldn't have been a question of trying to get it past the executive editor and editor in chief. I also knew there was no standard magazine practice dictating that an article accompanying a main article had to be like it in any obvious way: in fact, the more variety, the better.

And the threat of killing it was particularly galling, for women really matter in technopaganism; it wouldn't be honoring the reality of that special parallel universe to leave them out. That was the message I left on Battelle's voicemail: that *Wired* would look foolish, as if it had done a shoddy job, if it didn't talk about The Girls of Technopaganism. The story did run, but cut to 750 words.

So all of this, run-ins past and thwartings present, was eating at me the night of the Pagan Girl Massacre. I was gagging on the loss of face, full of self-loathing for having thought I was exempt from the second-class treatment other women had experienced at *Wired*, seething with impotent rage at knowing that if I no longer wrote for *Wired*, I'd lose access to *the* high-gloss, high-visibility, high-cachet venue for my writing, and most of all, wearily confronting one of Joan Didion's home truths. In her essay on "Self-Respect," Didion was talking about writers and Hollywood but her thesis rang true: it's a loser's game when you're trying to win the approval of people for whom you have contempt.

Where had I ever gotten the idea that being the house eccentric at *Wired*—its resident feminist/humanist/skeptic/ Luddite, the one they could point to as their woman writer, the writer who takes nothing at face value and refuses to cheerlead—could succeed? Silly, vain, foolish girl...

My *Wired* Problem, and Maybe Yours

Where things stand now, after getting much grief and much flack, is that the editorial boykings of *Wired* are finally promoting some women out of the pink-collar ghetto of back-office functions and support roles. A very credentialed woman was recently hired as a features editor. The research editor was recently made a features editor (meaning she can seek out and edit writers and story ideas on her own). Another woman was promoted to the position of assistant managing editor. Perhaps things really will change in the years to come.

But as I write now, I feel not very different from the very distinct minority of women I've met in high-tech who have made it there—and who quit in disgust after coming to what feels to them like their right minds and their senses. They come to feel that it's just not worth it, that it's a game not worth playing. They don't want to deal with the racket of male elephant-seal-like trumpeting and jostling for status.

If one is not interested in participating in circle-jerk exercises and paying homage to the alpha male of the moment, what is there left to do at *Wired?* All magazines engage in the happy talk of crowing about their subjects—unless their charters, like that of *The Atlantic Monthly*, are to be contrarian. In all fairness, I know it's my predisposition to ironic distance, my Petronius-Arbiter-of-Rome act, as much as my feminist malcontentions, that cause me trouble with *Wired.*

If I were more bluff and bully, if I had the heart and stomach and bile and spleen—not to mention the spirit and will—to keep slogging it out, in spite of knowing I do not belong, I would keep on writing for *Wired.* I would owe it to the principle of speaking truth to power and Not Wimping Out to assert that voice of agonized humanism and female snarkyness. But as with most folks of artistical temperament, I mostly want to be left alone to do my work, to get at that still silent voice within, and not be fighting ideological battles my editors at *Wired* might not even consciously realize they are engaging in. Progress means exchanging one set of problems for a more interesting set of problems—but my struggles with *Wired* have been generating problems of ever-decreasing interest. I strive to have other, better, more rewarding and less stupid battles to fight.

Being a token calls for tremendous carbohydrate-reserves, reserves ready to rush to the spirit-body's defense. And by 1995, in the field of publishing, where women have been present in large numbers for decades, it just seems as if this shouldn't be necessary; these are kilocalories best expended elsewhere: On the subject at hand. On saying things as best you can. On doing some small part to decrease the weight of human misery in the world or create good art. You shouldn't have to have the tirelessness of a revolutionary, or the sauciness of a turncoat, to be a woman and regularly write for *Wired.*

How Hard Can It Be?

Karen Coyle

> Louise: ... And steal Darryl's fishin' stuff.
> Thelma: Louise, I don't know how to fish.
> Louise: Well neither do I, Thelma, but Darryl does it—
> how hard can it be?
>
> —from *Thelma & Louise*, 1991

I Love the Smell of Silicon in the Morning

A woman says to me: "My daughters love science, and they use computers, but they just aren't interested in technology."

Try this one: "My daughter is an excellent driver, but she has no interest in internal combustion."

We're supposed to use computers, not worship them. There will be those fascinated with the machine qua machine, but we have no reason to assume that to be a superior approach. Except, of course, that it's the masculine approach by computer culture standards and, therefore, has the air of superiority. It's the difference between those of us who just want to get to where we're going and those of us who read *Car and Driver* magazine. Behind the wheel, we're all called drivers.

In 8.3 million households in America a woman is the primary home computer user.[1] Two out of every three on-the-job computer users are women.[2] But we still see computers as being a "guy thing." Stereotypes outweigh reality, as when Rosie the Riveter did a "man's job." How could it be a man's job if a woman was doing it? How can computers be masculine when women and girls use them every day?

But this masculine image is constantly reinforced in the computer culture and in the images presented in the consumer computing market. Rather than feminizing the computer field, women are obliged to adopt and maintain some degree of macho to become part of it. To question the masculinity of computers is tantamount to questioning our image of masculinity itself: Computers are power, and power, in our world, must be the realm of men.

Superheroes

PowerPC Pentium Power. Powerful Savings.
...powerful portable...
To understand the power of working together...
Project management software that is so powerful...
Powerful, but personal[3]

Selling computers is about selling power. Open any computer magazine and you'll find current cultural heroes, generally male sports figures, used in advertisements for all things digital. Some hero figures from the past are also employed, for example, Davy Crockett, frontiersman, standing high on a promontory looking out over the virgin territory before him. Even the anti-hero comes into play, as a lone figure in a Sam Spade raincoat advertises virus-detection software. Symantec sells its Norton Utilities with a Superman figure sporting the nerdish head of Peter Norton. Stacker's upgrade notices come emblazoned with "Stackerman," a hip, muscled super-hero.

Yes, in the mythology of the computer culture, computing requires great heroics and derring-do. Without heroes there would be no computers. It's no surprise, then, that Steven Levy's book on the early adopters of the culture of computer hacking is entitled *Hackers: Heroes of the Computer Revolution.* The key word here is "heroes," a decidedly male role with a great deal of power.

Levy researched his topic in great detail, especially the early computer activities at the Massachusetts Institute of

Technology, which began in the mid-fifties. He has a large cast of characters, all male, and even knows where they ate dinner and what their favorite dish was. In only one paragraph does he approach the question of why there were no female hackers, and answers:

> Even the substantial cultural bias against women getting into serious computing does not explain the utter lack of female hackers. "Cultural things are strong, but not *that* strong," Gosper [an MIT hacker of the early 1960s] would later conclude, attributing the phenomenon to genetic, or "hardware," differences. [4]

With little review of the facts, Levy also concludes that women are genetically unable to hack.[5] He never considers relevant that this hacking took place in a campus building between midnight and dawn in a world where women who are mugged at 2 a.m. returning from a friend's house are told: "What did you expect, being out at that hour?" Nor does he consider that this hacking began at a time when MIT had few women students. And though he describes his male hackers as socially inept, he doesn't inquire into their attitudes toward women and how those attitudes would shape the composition of the hacking "club."

But most of all, he never considers the possibility that among the bright women attending MIT at that time, none were truly interested in hacking. What if the thousands of hours of graveyard-shift amateur hacking weren't really the best way to get the job done? That would be unthinkable.

Levy's own adherence to the hero myth virtually excludes women from the field of study. Like those he studies, Levy finds women uninteresting. Roberta Williams, the woman who authored the first computer adventure games while her husband, Ken, ran the computing shop, is referred to only obliquely in the one-third of the book dedicated to the computer game company she and her husband ran. In Levy's account, Roberta is portrayed as a housewife and mother whose authorship of the popular games was the least important part of the process.

How little things have changed in the 150 years since Augusta Ada Lovelace, a brilliant mathematician who invented the concept of the algorithm, toiled almost anonymously in an intellectual world where only men were welcome. To provide her access to the library of the Royal Society in London so that she could continue her studies in mathematics, Ada's husband became a member. But as only men could enter the library, at one point Ada wrote to one of her male friends in the society:

> When Lovelace became a member of the Royal Society several years ago, it was entirely on my account. But the inconvenience of my not being able to go there to look at particular *Papers* &c which I want, continually renders the advantages I might derive quite nugatory.—Could you ask the Secretary if I might go in now & then (of a morning of course) to hunt out the things I require, being *censé* to do so in *L's* name....I suppose there are *certain* hours when I should not be liable to meet people there. Perhaps early in the mornings.[6]

Cultural bias *is* strong enough to give us a view of the world in which what men do is inherently important and the activities of women are only auxiliary. There are still those who claim that Ada did not write the mathematical pieces that appeared under her name, even though her own letters and notes prove otherwise. The bias is strong enough to determine when and where women can go out into the world and whether they will be accepted there, long before they have a chance to demonstrate their skills or prove their competence.

Although this shouldn't be seen as excusing women from heroic activity, it does mean that we can't look for women's contributions in male-defined institutions. It takes different eyes to see where women have been and what they have done. The role of women like Ada Lovelace or Admiral Grace Hopper (who was instrumental in the early development of assembler language) will not appear on the pages of books that look to glorify male heroes. If we accept the standard view of a male-dominated computer industry, we will never see the women who are making a contribution.

"Girlish GUI"

The machismo of computing survives, even though most computers are now "user-friendly" and no longer require a degree in engineering or an explicit interest in science or math. In a column in which he lauds the DOS operating system over the newer Windows OS, columnist John Dvorak says: "The original split between PC users and Mac users was a battle between the masculine command-line interface and the girlish GUI."[7]

Note that the graphical user interface is described as "girlish," not "feminine." The conflict here is between men and women, not masculine and feminine, since those are qualities that any of us can prefer, regardless of our sex. Or perhaps the conflict is between men and not-men; between the hacker types and all of the other people, both men and women, who prefer ease-of-use over control. If the machine itself is an instrument of power, then conquering it, controlling it, is like moving another step up in the evolutionary food chain.

Steven Levy attributes some of the early resistance to the Macintosh to the computer culture's concept of "macho" computing:

> The previous paradigm of computing—command-based, batch-processed, barely coherent—was deeply associated in the MIS [Management Information System] community with masculinity....Columnist John Dvorak contrasted the Mac with the new version of IBM's computer, the AT, and called the latter "a man's computer designed by *men* for men." [8]

Ten years later, the "girlish GUI" is still seen as an emasculating interface. After a lengthy round of posts on one Usenet group debating the command line versus the GUI, one participant replied: "I mean let's get really macho and demand the return of punched cards and core memory! ;-)"

Remember when we were eight years old and the boys all decided that girls were "icky?" In the computer field, we're still the example of "what not to be." Like Dvorak's "girlish GUI," femaleness often has a negative connotation in the computer world (unless, of course, it's in the form of a virtual girlfriend).

In this ad for a Microsoft multimedia CD about the composer Schubert, girls haven't gotten very far beyond icky.

"I'm in dis bar arguin' with dis jerk about Schubert. I sez to him, 'The essential Schubertian style is in the unfolding of long melodies both brusque and leisurely. That's the blessed earmark of Schubert's style and it's all anyone needs to sense his distinctive attitude toward musical structures.' Well, in a high girlie voice, the jerk tells me, 'By classical standards, it's fairly loose construction.' To which I replies, 'Dis is Schubert, tough guy; and belaboring him with his musical ancestors is like comparing apples to oranges.' And then I decked him."

The "high girlie voice" got this guy "decked." The speaker is featured as a big, hairy biker type. Not the image of your typical male computer user, but clearly heavy on the desired testosterone scale. In this case, the negative image of femaleness also translates to a more than plausible homophobia.

"Friendly to Your Hand. Deadly to Your Enemy."

They're called joysticks. That should be the first clue. At the computer store, they are lined up along a shelf, each black, erect, six to eight inches tall. An improvement on nature's original design, they have neat grooves that fit your hand so you can hold them for hours as you use them to kill and maim while playing computer games. But I just can't bring myself to reach up and grasp one in public. It seems so...well, penile, and I feel shyly inexperienced.

Each has a button on top that can be fired, again and again. One carries the logo "Use It or Die." How biblical! Here we have the high-tech version of Norman Mailer's "killer penis," which was a shockingly macho fantasy even for Mailer. Maybe we should name one of these joysticks after Mailer, though I suppose he secretly wishes he were named after one of them: *The Thrustmaster*.

Penis as weapon; weapon as penis. Do we fight back with a virtual *vagina dentata*? My local sex shop has quite an array of flexible, woman-friendly silicon dildos in cheery purple and pink swirl designs. This is an entirely different use for silicon—but is it hardware or software?

Computer games are notorious for their exhibition of macho. Most games are of the "if it moves, shoot it" variety. Like male-oriented pornography, the games get right into the action with little need for plot, motivation or character building. The character that represents the gamer is often a hulking male figure. There are few female characters, unless they need to be rescued from trolls or other such monsters (including other men who look much like the gamer character). The common wisdom is that these games appeal overwhelmingly to men and boys, while women and girls are attracted to games like Tetris or Solitaire.

> One game advertisement gives away the secret:
> The average male reaches his sexual peak at age 17.
> And lives to the age of 73.
> So what do you do with the 56 years in-between? [9]

The computer game is a form of virtual virility. The themes are hyper-macho and often revolve around the activities of sports and war. They come with names like Hardball 4, Front Lines, Body Count. The ads for Doom, 1995's most popular game, read: "Now there's a place more violent than earth."

I get a mental picture of paunchy nerds or pimply-faced teenage boys, able to convince themselves they really are the muscle-bound hero of a game called Mortal Kombat. It angers me that we encourage this self-glorification in men, while at the same time the message we give to girls is that they don't measure up. Measure up to what? The kind of violent insanity we expect of men?

What would games be like if we designed them with a female audience in mind? Would they be like the Barbie computer game where Barbie gets new outfits and learns to be a fashion model? Or could we conceive of a game where a clever woman saves the world for all humankind? Unfortunately, even our fantasies for women are based on lowered expectations.

Then again, we women reach our sexual peak much later in life and maintain it longer. Mentioning this in public is taboo; mentioning it in private gets women beaten up. The system is perpetuated by our unspoken agreement to participate in this fantasy of male superiority. Computers are ideal partners in this vision. They are the soulless companions that we women are unable to be—obedient and unquestioning.

. . .

It's a Guy Thing

The man sitting at the desk in the printer ad holds a basket-ball in one hand. Another ad has a basketball with "Intel Inside" embossed on it. Under the words "Internet Made Easy" an arm poises to slam-dunk a basketball-sized world.

All these basketballs, and not one Nike shoe ad in the whole magazine. This isn't about sports. If you want sports, you read *Sports Illustrated* or the sports section in the newspaper. This is about being "guys." It's about balls. Big balls. These ads have little to do with computing or sports, and everything to do with the image of maleness. And where there are males, there are powerful cars, or at least car metaphors:

> Vroom! Digital Engines Put a Pedal to the Metal "These days, though, Holley carbs, horsepower and cubic inches are passé. The real speed thrills are to be found roaring down the info highway, where the measure of muscle is MIPS...and bandwidth..."[10]

On their Technology & Media page, the *New York Times* ran an article by John Markoff, excerpted above, on the "state of the art" in computing that was written entirely in hot rod metaphor. The article cleverly presented computing as the successor to the typical male bonding activity of car talk. When the author was approached (via email) by a group of women, he was surprised to hear that his hot rod theme was viewed as "masculinizing" the computer. His reply was that as far as he was concerned, women and men are equally interested in cars and speed.

But it takes only a brief glance at magazines targeting men and those targeting women to discern how different the themes are in these publications. If one were to believe *Sports Illustrated* and *Women's Day*, men drink and drive (though on different pages), and women cook and diet. Neither seems to be a fully representative or healthy view of the sexes, but it is striking that there is virtually no overlap in content or in advertising. In his article, Markoff has only given voice to a divide abundantly clear on the page.

Undoubtedly the *New York Times* believes that its content is gender-neutral. But as is so often the case, neutrality is much more masculine than it is feminine, especially in discussions of technology.

It's no wonder that women have a hard time seeing

themselves in the computer culture. Advertisements that talk to "you" almost invariably feature men, making it clear that women are not the audience for computer wares. Review copy talks about "pictures of your wife and kids" as if it were inconceivable that the reader be anything but male (and heterosexual).

The exclusion of women and femininity is as obvious as the inclusion of male imagery. And when women do appear, it is often as objects of desire. As if the more subtle ways that computer culture excludes women were not enough, the frequent presence of pornography in the computer culture alienates many of us as well.

The image of the nerd is male, socially awkward and sexually frustrated. And as a matter of fact, the computer industry is an active purveyor of porn. Computer trade shows are one of the few "professional" situations today where women are confronted with pornography being displayed and sold openly. At a recent MacWorld exhibit, two booths sold CD-ROMs with such titles as *Anal ROM*, with the actresses present to sign copies.

CyberGirls

The back pages of many consumer computer magazines are taken up with ads for a variety of "sexy" offerings most likely to appeal to men: Some of the ads below came from the back pages of *PC Magazine*, while others are from the free advertising-based computer magazines *Computer Currents* and *MicroTimes*.

> *Screw* Magazine on CD-ROM: HOT! HIP! RAUNCHY! OUTRAGEOUS!
> The 'Infamous' GaRBaGe DuMP: Caution, Adults at Play®
> Grafix: Computer pictures so hot they could melt your modem!
> Throbnet
> Let Us Blow Your Hardware!

It's hard for me to imagine the appeal of looking at dirty pictures on a grainy, 14-inch screen when I can take a full-color magazine to bed with me for my enjoyment. And it's pretty hard to think of sitting at my desk in my straight-backed office chair getting sexually excited. Maybe this is an area where biology really does determine a difference between male and female behavior, because essentially I'm, well, I'm sitting on it, so this position just doesn't work for me.

> Now You Can Have Your Own GIRLFRIEND™...a sensual woman living in your computer! First VIRTUAL WOMAN. You can watch her, talk to her, ask her questions, and relate to her. Over 100 actual VGA photographs allow you to see your girlfriend as you ask her to wear different outfits and guide her into different sexual activities. An artificial intelligence program with a 3,000 word vocabulary that GROWS the more you use it. She will remember your name, your likes and dislikes.[11]

Finally, a woman he can relate to, and one who even remembers his name. (And I bet he wishes that it did GROW the more he used it.) This is the "artificial intelligence" version of the plastic blow-up doll. A full relationship without having to involve another human being.

It's pretty clear what the market will be for virtual reality. The first "virtual sex" is already on the market.

> Hi. I'm Girlfriend Maria.™
> May I come live in your computer?
> You won't believe your eyes (or ears) when this new Girlfriend takes up residence in your PC! She's cute, she loves talking to you, and she actually learns! Peek in on her as she goes about her daily activities, eating, playing the piano, watching TV or reading. And she's *really* interactive; no archaic, limited mouse menus— she responds to anything you say! This is the revolutionary *artificial intelligence* game everyone's talking about, so get your copy now! Rated G, so it's safe for the kids, too![12]

Well, at least she can play the piano. I mean, that shows she's had some proper upbringing. Not like some of the girls, the ones with the questionable backgrounds. And she's clean, so you can even introduce her to the kids without any embarrassment.

I can imagine Maria. Maybe she hasn't been in the country for very long. She's just trying to make a better life for herself on this side of the border. And if that means keeping some guy happy, well hey, as long as she doesn't have to do anything...you know.

And she learns, she learns really fast. Next thing he knows, she's popping up with little suggestions: "Move this out to three decimal places then round off—you'll find it gives you better results" or "A slightly bolder font, perhaps Garamendi, will give this document a more forceful look." He turns to her with more and more questions, then finally just turns the computer over to her. She's much better at it all. Only, one day he notices something funny about his credit card bill...and his hard disk is missing a lot of files...and Maria's gone without a trace, except for the mysterious message that appears at random intervals on his screen: "Hasta la vista, baby."

Women as Roadkill on the Information SuperHighway

Women are a minority on the Internet today, at least as active participants. Although some statistics show that as many as 30 percent of Internet accounts are held by women, they're not necessarily taking up a third of the bandwidth. And it's certainly obvious to women who make their first forays into cyberspace that it's mainly men out there.

The highway metaphor lends itself well to masculine images. The same *New York Times* article that talked of computers in hot rod lingo showed a cool dude with a souped-up computer leaving tire tracks over his slower rivals. To the side, the only woman in the picture (dressed in fifties bobby-soxer style) looked on admiringly with the caption "Ooh, what MIPS!" It's not that long ago that in many activities women merely cheered men on, but it's hard to accept that image applied to the technology of the future. That's not the future we've fought for.

An advertisement for software that allows a direct Internet hookup from a personal computer shows a man on a motorcycle and the caption "Pop a wheelie on the Information Highway." On closer inspection you see a pair of high heels flying off the back of the bike: His wheelie has just dumped a woman on the road. Not a friendly image for women, but the guy is portrayed as having the time of his life. If the company saw women as potential customers, it would not—could not—have chosen that image: woman as roadkill.

Another illustration shows a more subtle type of masculine appeal. The cover of a local weekly highlighted an article entitled "Plugging In: An Idiot's Guide to the Internet." The background is a circuit board with heavily made-up, sexy female eyes. In the center (about where her nose would have been) is the screen of a laptop. Into the empty space of the screen an enlarged phone cord is "plugging in." You have to wonder if the creators of these pieces are aware of the sexual message of their imagery. In this case, I'm not sure. But the appeal of the Internet is presented as sexual, or at least sexiness, and the intended audience is a man—even if this particular man is also an "idiot."

How Hard Can It Be?

The assumption in our society is that men's activities are difficult, and that is why women can't or don't engage in them. Women's activities are, of course, inferior, which is why men don't engage in them. So when we look at a man fishing and a woman quilting, we make some assumptions about the skill or strength required for the task. We need to start looking at this in a different way. If men choose to spend their time playing golf or waxing their cars and women choose to spend that same time making dessert or cleaning closets, does this really tell us anything about the amount of skill required or the kind of respect we should give to those who perform these activities?

The difference isn't in skill but in the social status already assigned to the activity, and we all support this status with our assumptions of the superiority of male activities. What needs to change is not the activities of women and men, but our attitudes toward them.

Men are unlikely to be supporters of such a change in consciousness. For women to become full participants in the computer field and in the use of computers, we are going to have to make a lot of revisions in our concept of machines and power. One possibility is that the computer will eventually be seen as a simple appliance, like the refrigerator. When the computer is stripped of its image of power, it will become acceptable for women to be computer users.

But it is doubtful we will be able to demote the computer to appliance status in the near future. The inevitable march toward the development of the "information superhighway" means that we are depending on the power and mystique of computers to provide new markets for our economy for the foreseeable future. And a machine with all of the fascination of a toaster won't motivate the consumer market.

No, what we will need is a conspiracy of sisters that begins with the recognition that there is nothing inherently masculine about computers. We must learn to read the computer culture for the social myth that it is. And we have to teach our younger generation of women that they are free to explore computers in their own way and to draw their own conclusions about the usefulness of these machines.

And we start it all with a simple thought that could be the beginning of a revolution: *How hard can it be?*

Endnotes

1. Rachel Kaplan, "The Gender Gap at the PC Keyboard," *American Demographics*, 16 (Jan. 1994): 18.

2. Kyle Pope, "High-Tech Marketers Try to Attract Women Without Causing Offense," *Wall Street Journal,* 18 March 1994, B1.

3. These phrases are all taken from computer magazine ads, mainly from *PC Magazine* in 1994. Any one issue can yield this many examples and more. And a Finder search for "power" on my presumably less power-oriented Macintosh yields: *PowerPoint, PowerPort, PowerModem, Power Tools, PowerUsers.*

4. Steven Levy, *Hackers: Heroes of the Computer Revolution* (New York: Dell, 1984), 84.

5. George Gilder, of the Discovery Institute, comes to the same conclusions. He equates male domination in the computer field with their same dominance in mathematics, logic, politics, physics, chess, athletics and violent crime ("Women and Computing: Nature or Nurture?" *Los Angeles Times*, 11 April 1994, S16). (No mention is made of the correlation between logic, computer use and violent crime; as a matter of fact, the latter rarely seems to coincide with the other two.) Gilder cites biological differences to explain why women aren't competitive with men in the world of computing.

6. Betty A. Toole, *Ada, the Enchantress of Numbers: A Selection from the Letters of Lord Byron's Daughter and Her Description of the First Computer.* (Mill Valley, Calif.: Strawberry Press, 1992), 305.

7. John Dvorak, "DOS is Alive, and Well...," *PC Magazine*, 13 Dec. 1995, 3.

8. Steven Levy, *Insanely Great: The Life and Times of the Macintosh, the Computer That Changed Everything* (New York: Penguin Books, 1994), 197.

9. Advertisement for 3DO game machines, in *Electronic Gaming Monthly*, 65, Dec. 1994.

10. John Markoff, "Vroom! Digital Engines Put a Pedal to the Metal," *New York Times*, 3 Jan. 1995, C18. Notes that MIPS is Millions of Instructions Per Second, one of the various measures of the speed of a computer.

11. Advertisement for "Sexy Software" which appeared in several issues of *PC Magazine* in 1994.

12. AIVR Corporations (Richardson, Texas), *Computer Gaming World*, Dec. 1994.

Closure Was Never a Goal
in this Piece

Judy Malloy and Cathy Marshall

In 1993, Xerox Palo Alto Research Center began a new program, PAIR, the PARC Artist In Residence program. The program was created to bring together scientists and artists, with the hope of initiating a dialog between the two communities, and creating what PAIR program director Rich Gold described as "new art" and "new research." We were one of four initial artist/researcher pairings, and in fact, Judy Malloy was the first artist to be in residence at Xerox PARC. We were also the only two women paired—an unsurprising anomaly, considering the communities that we belong to. In this essay, we discuss our collaboration, the work we produced together, and more tangentially, the eerie machinations of fate during the year and a half that we have worked together so far.

We have written this essay in a manner similar to how we produced the work we discuss: each segment of the paper is an author's response to one (or more) of the preceding segments. Like the work itself, the structure of the essay is not like a tennis match—multiple associations may stem from one source, a given association may draw from many sources, and the associations may be oblique. The notable difference between the two is that the essay segments are longer than what we refer to as the "screens" we exchanged as part of the actual piece. We hope that the use of this technique in the essay will give you, the reader, a feeling for the process. We have titled each of the short sections the way we would have tagged our screens, and refer to individual screens quoted from the work itself by their number and subject line.

Judy: A Collaboration Seemed Unlikely

The day I first met Cathy I was menstruating heavily and was worried that the blood would leak through the Kotex. To make matters worse, the frozen peas I sometimes use to deaden back pain during my period had thawed and were leaking pea juice down my legs.

It was not a good day for Cathy, either, because the people she referred to as her "funders" were there. Nevertheless she showed me a hypertext document-authoring tool she was working on. I particularly liked the way color was used to distinguish types of screens.

I saw her later in the hallway, flanked by several important-looking people in suits—"funders"! At the time, a collaboration between someone surrounded by an aura of blood and pea juice and someone flanked by funders seemed unlikely.

Cathy: From Corporate Offices to Sunny Backyards

Judy's and my initial meeting seemed like a hurried appointment—too corporate, too metal-and-glass. I gave her a quick demonstration of our system, talking fast, explaining little. Then we moved on to what is referred to by researchers at Xerox Palo Alto Research Center (PARC) as the "beanbag room," a reminder of the center's roots in the seventies, for Judy's artist's talk.

Judy read excerpts of past work off a series of accordion-folded, taped-together, three-by-five-inch cards, a technique that made a strong impression on me (I even tried it in a subsequent technical talk). By the end of her artist's talk—especially after hearing Judy read several oddly familiar narratives about flashers[1]—I knew I wanted to strike up a collaboration, but was nervous about whether we could transcend our related roles within the bounds of the PARC-sponsored program, or that she'd even want to work with me after our inauspicious first meeting.

Is masquerading as a responsible adult, a scientist, a researcher, a businessperson, a form of transvestism? I thought about this, men in suits in tow, as I glimpsed Judy in conversation with another researcher later in the day. We all must do this, I thought.

Our second meeting took place in the garden behind Judy's basement apartment in Berkeley. I was an hour-and-a-half late and once again felt very awkward. This meeting grew more relaxed as we sat in the sun. Tess, Judy's cat—the one with the wild yellow eyes—was there, too. We talked a little about MUDs and MOOs[2] and about living and working in basements, something we had in common.

Judy showed me an artist's book she had done as a cardfile. The cards talked about itching. I liked the tone of it and hoped we could strike a similar tone in our collaboration, although I didn't say so at the time.

That day, we decided to work via email and worry about details of structure and organization later. The topic of individual pieces would grow out of this conversation in Judy's sunny backyard.

Judy: It Was Surprisingly Easy to Talk

I was holed up in my basement on the Albany-Berkeley border, scraping out a living on the fringes of the Internet—with a broken heart, an extraordinary black cat and no car. Xerox PARC seemed so far away that it might have been on the moon. So Cathy's call and willingness to drive what seemed to me at the time an immense distance was an unlikely miracle.

I am finding it impossible to visualize us sitting outside that Sunday on white plastic chairs without seeing two very different things—my crutches leaning against the side of the house and electronic mail from Mark Bernstein.

No. The crutches were not there then. I have been on them for so long that when I remember Cathy's reassuring blue jeans and the coffee and beer on the table with the faded oilcloth cover, I wonder how I managed to get the coffee to the table. (Beer cans are easily transported in pockets or if necessary thrown through the door so they land close to if not always exactly on the table.)

If it hadn't been for Mark Bernstein, the integrator who somehow brought hyperfiction/hypertext authors together in the Eastgate Systems fold,[5] we might not have been sitting there talking. Mark had thought we would work well together. At least we should talk, he had said.

It was surprisingly easy to talk. From the beginning the project evolved seamlessly and collaboratively. The work would be an exchange of details about our lives—like those conversations that occur in coffeehouses or bars between people meeting for the first time. By electronic mail we would look for the links in our artist/researcher existences—basements, brothers or sons studying history, the assuming of personas, the playing of multiple roles.

Cathy: A Postcard Suggesting Fiction

What I visualize from our first two meetings—besides basements and gardens, beer and coffee, offices, meeting rooms and the Pod 26 restroom—is a postcard, a duplicate of the one I had pinned up on my office bulletin board when I first received it, earlier that year.

The postcard is from Eastgate Systems and promises "WOMEN, HYPERTEXT, AND ART: important new hypertext fiction from Eastgate." It is an announcement of two works of hyperfiction, including Judy's "its name was Penelope." The postcard I remember so vividly is annotated with Judy's email address on the WELL[4] and directions to her Berkeley basement (in a subsequent penciled scrawl). Its twin has faded, for the bulletin board hung beside a sunny window that looked out onto the live oaks and horse pasture next to Xerox PARC.

Fiction. The postcard made me think "fiction" before our first meeting. But, during our second meeting, we decided on a course that avoided unnecessary invention—to exchange the real remembered substance of our lives. The links would arise naturally within this constraint, shaping the work in an organic way.

We agreed, at our meeting in Berkeley, to define structure later in the project, after we had amassed content by email. The links were to be left more or less implicit in our exchange (although, in practice, associations often found their way into the Subject line).

In retrospect, email seems like a naturally hypertextual form, with its splitting and merging threads of conversation, its subjects that recur and re-emerge and its tendency to discourage linearity and closure. Of course, I didn't really think about this in Judy's garden in Berkeley. The sun, for me, is a narcotic.

Judy: A Series of Fires

The assumption of personas is integral to performance art, so I am accustomed to distorting experience into fiction in the first person—to speaking as an "I" that is not really me, but, rather, some other woman whose voice I have assumed for the duration of the work in which she speaks. I anticipated that this unmasking, this speaking as myself might be uncomfortable, but to the contrary, as our work began by electronic mail between *well.sf.ca.us* and *parc.xerox.com* in late September of 1993, I found that speaking as the real "I" was an empowering experience.

From the very beginning, although Cathy's enviable Southern California childhood

(cathyf.001: no basements) My friends and I were curiously passive, without overt ambition. When our teachers would ask us why we hadn't been in class, we'd shrug and say, "Went to the beach," as if full disclosure would suffice.

seemed worlds away from my colder New England memories,

(judyf.002: school was inevitable) After dark, as I lay under several layers of wool blankets, night trains periodically rattled the storm windows beside my bed. Over and over I dreamed that I was being chased by a relentless non-track bound freight train that was able to follow me no matter in which direction I ran.

there were unifying themes. A series of fires, for instance:

(judyf.002: basements) For years after the newspapers caught fire in the basement of the house in Winchester, my mother made sure that the only matches in the house were safety matches. She thought that somehow the hot weather and some strike-anywhere matches in the immediate vicinity of the pile of old newspapers had resulted in spontaneous combustion. Actually, my brother, whose chemistry set dominated that area of the basement, had lit the newspapers on fire just to see what would happen.

(cathyf.002: failed pyromania) I would've liked to have been a pyromaniac like all the other kids; with all that dry brush around, I really could've made a splash. Set one of those roaring hillside brushfires that stings your nostrils and makes the LA sky orange even at midday. People all the way down in Torrance would've had to comb bits of gray ash from their hair.

As it was, the first time I tried to light a match, I failed. Miserably. Not enough friction, I guess. Scrape. Scrape. You could smell the phosphorus. Scrape. I was supposed to be doing a science experiment in front of my fourth grade class. Scrape. They were all watching, waiting for me to light the match. It was simple: One side of a tin can was painted black, the other side white; a candle in the middle would heat them up. One side would become warmer than the other. But the match wouldn't light.

My father burned down his house when he was five by playing with matches in the closet. His father was the Fire Chief. His grandfather was the Chief Dispatcher.

I was a failure. My match wouldn't light.

Years later, Tom told me how he fought a fire in the East Bay Hills when he was a teenager high on PCP. How the wind fanned the fire in every direction. How the flames rolled down the slope, and how he finally ran.

Judy: The Blending of Two Real Lives

Rich Gold, the director of PAIR (PARC Artists in Residence) had presented the program to me in terms of building bridges between artists and researchers. So, although I usually write fiction, it was this bridging of our lives as artist and researcher that was on my mind.

Also, fictional places, things and people have a way of acquiring a life of their own. The manipulation of "my" characters that sometimes occurs in collaborative works can be uncomfortable for me. An alternative is to intertwine stories the way Guyer and Petri did in their collaboratively written hypertext, Izme Pass,[5] but they knew each other and each other's writing styles. We didn't, and I had no idea how our voices would merge.

What was obvious was that we had both led full and

interesting lives. Or to put it another way, the strength of Cathy's life was made immediately apparent by the obvious rarity of women researchers at PARC, but because I knew nothing about her writing, I thought in terms of discovery and exploration. In retrospect, as is obvious from Cathy's writing in our work and the way our voices cohere, we could have written intertwined fictions. Nevertheless, what we did—the blending of two real lives—seems more extraordinary. Rereading the mystery-filled unfolding of our oddly linked lives still sometimes sends chills down my spine.

Cathy: The Fluid Structure of Time

Time became very fluid, malleable, during our exchanges. We'd flit back and forth between childhood, earlier in the day and indeterminate points in between. Chronology of the events in our exchange is elusive and reclaimed only by the construction process that each of us plays as reader. Four excerpts from screens that use the ocean as a backdrop span childhood, a night camping on the beach sometime in the late 1970s, the ocean imagined from the writer's current view of the desert, and an eerie dream after Judy's accident in Phoenix.

(cathyf.120: tidepools) ...chitons clinging to rocks, darting sculpins, ocean-green anemones (their tentacles swaying with the mild swells).... Perfect worlds, miniature, bright and moving.

(cathyf.029: narrow bridge) The ocean grew louder. A million chirping frog voices merged with crashing waves... "I think I'm stepping on frogs, man." Corky's voice floated back toward us.

(judyf.097: oasis) Here there is no cold, deep blue-green ocean water, no jellyfish drying in the sun, no tide pools in the beach sheltering rocks.

(judyf.104: bridge) We are way up in the orange scaffolding, moving slowly, helping each other. The Pacific Ocean is black beneath us.

The screens grew to be a collage of (sometimes overlapping) memories. A brief screen about a waterbed seemed oddly familiar when I read it during a later editing pass. I mistook it for one of my own.

(judyf.042: blue bed) In Albuquerque, our blue plastic waterbed was translucent so that you could see the water rolling around and the mold growing on the inside.

I like beds to be solid under my body—like floors.

What, I wondered, was I doing in Albuquerque?

Cathy: The Reader As a Third Voice

Because so many of the associations between our screens grew out of particular words, we decided to include a generate function[6] in the final work. Judy first used a technique of this sort to gather excerpts from our collected screens for a talk at Stanford. She used a search command to collect all the lines from our screens that referred to beer and create a composite screen, which she then edited for presentation.

This, we reasoned, was something any reader might want to do—request a screen to be generated from the work based on a thematic word. In the final work, the "lines" button first requests a word from the reader and then gathers the lines containing that word. Prior to display, each line is automatically marked up so that it is linked to the screen from which it originated. In this way, the reader might request the work to generate a screen about, say, roaches. In the example below, some lines have been omitted for the sake of brevity and the link anchors are not represented; but the excerpt captures the essence of the results:

A big cockroach picked his way along the baseboard
He knew the ins and outs of the roach motel.
We were used to the roaches. Roommate deplored them,
The roaches thought he was a four-star chef
boric acid, and multipurpose cockroach death chemicals
Cockroach discussions with Mr. T., who was my landlord
In the pool, a big cockroach was swimming frantically for shore.
cockroaches rustled in the sink at midnight
cockroaches the size of small mice

This technique adds a third voice to the work—that of the reader, since it is the reader's choice of words that drives the gathering process. I have been surprised in my tests to find that even arbitrary (nonthematic) word choices like colors lead to interesting automatically generated screens.

Judy: At Xerox PARC

In November 1993, I was at Xerox PARC—working in the Computer Science Laboratory (CSL) with Pavel Curtis on a different project (a narrative in LambdaMOO called "Brown House Kitchen"[7]). Somewhat ill at ease, at a luxurious SPARC workstation,[8] in a temporary office, I was reassured by Cathy's responsive screens. On the difficulty of conversing with fellow researchers, for instance:

(judyf.058: a stupid question) Tried to think of things to say. I desperately wanted to communicate. Stuck carrots and mushrooms into my mouth. Fourth tea I had been to, and it was only getting worse. I ate a fudge brownie. "Where does the food come from?" I asked. Everyone laughed. "Elves," someone said, and I relaxed for a minute, but they moved away.

(cathyf.072: celestial small-talk) We were leaving the building at sunset: My colleague pointed straight out ahead of him and asked me, "Is that the sun or the moon?" The moon was hanging right on the eastern horizon, enormous, white, almost luminous. It might've looked a little like the sun.

"The moon," I said, unable to come up with a more appropriate rejoinder, and we strolled the rest of the way to the parking lot in silence.

Soon after I began working at PARC, Cathy and I had lunch in the cafeteria. Questions and answers flowed quickly, and it was immediately apparent that meeting face-to-face would dilute our exchanges. As a result, reluctantly for me at least because conversation with Cathy was a welcome oasis from the heady, fascinating but code-focused, mostly male, monastery-like CSL, we agreed not to schedule face-to-face meetings until the following August when we were scheduled to spend a month together designing and programming the work we were creating.

Sometimes, however, I waved when I passed her window on my way to PARC in the morning. I took a route that Cynthia Duval, an ethnographer working with the PAIR program, had shown me that involved cutting through the field with the horses, ducking under the barbed wire and hiking down the small, steep hill that Cathy's window overlooked. On the way I picked up things from the field, which I deposited on the shelves of my office—dry grass, desiccated horse shit, horse chestnuts. The horse chestnut theme this began is another area of the work where sometimes it is difficult to know which one of us is talking.

(judyf.052: horse chestnuts) or took them home and laid them out on the table beside my bed. Every night, while I was sleeping, they grew dull and wrinkled.

(cathyf.063: horse chestnuts) The older girl and I ran off the trail to gather the horse chestnuts, smooth as river stones; our pockets bulged with them, yet we kept gathering, giving them to the others to carry.

(judyf.053: horse chestnuts) I lay them in neat rows across the top of my workstation monitor or throw them into the tall grass in the fields where the horses are.

(cathyf.064: horse chestnuts deflated) The horse chestnuts are in a clear plastic box on my desk. They haven't collapsed too much more than they had when I retrieved them from the truck and brought them into the house.

Judy: Another Accident in Phoenix

I am lying here in Berkeley, looking out the basement window. Some of my words in Cathy's and my collaboration were written in Berkeley, some in Palo Alto, some in Arizona where I was doing some work for Arts Wire[9] in the first half of 1994. Now I am back in the basement in Berkeley where it began. Cathy's surroundings have also changed during the year and a half we have been working together. Now, she is writing me from College Station, Texas.

It is raining, and in the damp air, the metal rods in my leg which weren't there when this work began, persistently ache as I review the trail of accidents and scars in our work that (so fittingly from a literary sense) ended with "another accident in Phoenix." Like the fires, these accidents and catalogues of scars began early.

(cathyf.006: a catalogue of scars) Mark has scars, too, one on the bridge of his nose where a cop hit him hard with a closed fist when he was wearing glasses. He had been riding his motorcycle: It was a motorcycle accident of sorts.

(judyf.008: internal scars) I remember seeing the wheels of the ongoing cars as my head skidded on Grove St. after the kid in the station wagon ran into my motorbike. I remember that a man on the sidewalk told me that they thought I was dead, and that while I lay there unconscious for several minutes, nobody had stopped the rush-hour traffic.

(judyf.005) When his brakes failed, Jim was following me home on the motorcycle, so it was my car that he slammed into.

(cathyf.023: another accident) Yesterday's accident in Phoenix reminds me of Corky. After the crash, he could only move the muscles above his shoulders. His neck was knotted with tension as he watched The Black Stallion on the VCR in his room at the nursing home.

(cathyf.024: he called his ex-wife) His ex-wife was more sanguine than he expected. Calm. We won't know for a few days, she told him.

She was already making plans to help her son learn to walk again. First she'd teach him to crawl, like she did when he was a baby.

(cathyf.034: mid-conversation) Mid-conversation today, my mother asked me, "How's that boy doing?"

"Corky? Mom, he died more than a year ago, in the summer. I remember telling you." I was standing in front of the kitchen sink when I said this to her, because it seemed just a little warmer there. I had abandoned the crossword puzzle at the kitchen table.

"That's right," she said, "I remember. Drug overdose?"

"It was a car accident." I didn't want to talk about it a second time.

Nor did I say anything about the accident outside of Phoenix almost a week ago, even though she'd known both boys.

In July 1994, about half a year after Cathy sent me the series of screens about the accident near Phoenix, I was crossing Mill Avenue in Tempe, Arizona when a pregnant seventeen-year-old girl with no driver's license was looking at the shopping center instead of the road and hit me.

(judyf.102: another accident in Phoenix) I don't remember what happened. The witnesses tell different stories:
"She was walking a bike on the crosswalk."
"She was riding a bicycle across Mill Avenue."
"All of a sudden a car appeared and struck her."
"I heard a scream."
"The victim was an older woman on a bicycle."

"The victim was a girl."

"The accident victim flew through the air along with her bike."

"Saw her fly through the air and fall."

I remember what looked like a disconnected rubber foot, bleeding profusely with bones protruding, lying beside me on the unbearably hot pavement.

I spent the rest of July at Maricopa County Hospital in Phoenix where a series of surgeons bypassed my severed artery with pieces of vein "harvested" from my ankle, inserted rods and screws from my hip to my ankle and finally covered the exposed bones with pieces of muscle and skin from "donor sites" on other parts of my body. Despite the reconstruction, there was an 85 percent chance of losing my leg.

In August I emerged from the hospital. I had acquired a fresh patchwork of new scars, my leg was intact, but there was no way I could go to Palo Alto as originally planned and work with Cathy. Instead, I lay in bed—alone in a temporary apartment in Arizona.

We talked on the telephone. We are both interested in virtual communities, virtual collaboration, so we decided that a virtual way of continuing and finishing our work was not inappropriate and might even be interesting. We agreed I would begin the process of organizing the work by separating the screens from their email headers. Then I would number them and upload them to a jointly accessible UNIX account. Later, Cathy, whose programming skills are more advanced than mine, would do more of the actual programming work.

In bed in Arizona with a Xerox PARC laptop I broke up our screens, separated the mail headers and edited screens—sometimes working while visiting nurses changed the dressings. The way my leg looked, the pain that made me at times suicidal, the frightening, solitary transferring from bed to wheelchair to get to the kitchen and bathroom—these were temporarily forgotten as I worked.

Cathy: Beginnings, Endings and All That Lies Between

It has been over a year and a half since Judy and I met over thawing peas and funders in Palo Alto and over beer and coffee outside her basement apartment in Berkeley. Her screens about her accident in Phoenix mark the end of the first phase of our collaboration on content. Since then, we have been focused on structure—the details of how to articulate the connections, how to present the individual screens we produced during the course of our project, and how to give a reader some sense of the process itself.

Structuring the work has been more difficult than I expected. The fluidity of the process obscured the complexity of the structure. It is both densely interconnected and loosely woven. Should we expect a reader to experience the screens in the same order in which we wrote them? Should we put the reader in front of a CRT in a darkened motel room?

(judyf.046: black vinyl headboard) I have the lights turned out. Yellow words emerging from the black monitor.

Should we choose multiple beginnings and multiple endings? Should there be coherent threads through our lives? Chronologies?

Besides adding the gathering function described earlier, we also decided to include a random function that brings a reader to a screen selected at random from our entire collection. The random function addresses the high inter-connectivity just as surely as using a large number of explicit links, since the effect is in some ways quite similar—a reader can get to any screen from any screen. It is the forward and backward functions—and other explicit links—that have given us pause. It is perhaps the sum of the experiences of past screens, the cumulative mystery, that has led us to write the next. How can we present this in the work itself?

We have continued to exchange screens as we structure the results of the first phase.[10] I was delighted when Judy suggested that we continue beyond our original time limit, keeping the work going indefinitely. As I write this, I still look forward to seeing email from *jmalloy@well.com.*

Closure was never a goal of this piece.

Endnotes

1. The two narratives Judy read are from her work, THE YELLOW BOWL:

He approached the tent where I was sitting sewing
—a middle-aged man with brittle brown hair.
He opened his raincoat, under which,
like some newspaper story flasher,
he had nothing on.
Holding his hand on his erect penis and gesturing with it
in my direction, he said:
"Voulez-vous Mademoiselle?"

The man opened his white lab coat
and drew a shiny orange red mushroom with rough white spots
and brilliant white stem
from an inside pocket.
He held it in his hand gesturing with it in Helen's direction.
Do you want some, he asked Helen politely.

2. MUDs (multi-user dungeons) and MOOs (MUDS Object-Oriented) are publicly accessible text-based virtual reality systems. For further information, Pavel Curtis, "MUDding: Social Phenomena in Text-Based Virtual Reality," *Xerox PARC CSL*-92-4, April 1992:21.

3. Mark Bernstein is the chief scientist at Eastgate Systems, a publisher of hyperfiction and hypertexts as well as of Storyspace, a hypertext authoring system. For more information, see http://www.eastgate.com

4. The WELL (Whole Earth 'Lectronic Link) is a computer conferencing system based in Sausalito, California. For more information, send mail to info@well.com

5. Carolyn Guyer and Martha Petry, "Notes for Izme Pass Expose," *Writing on the Edge* 2(2):82-89, Spring 1991.

6. The function, implemented as a clickable button, automatically gathers single-line excerpts from the work.

7. Created in collaboration with Pavel Curtis, as part of my 1994 residency at Xerox PARC, the narrative BROWN HOUSE KITCHEN is disclosed by the objects in a room of that name in LambdaMOO. Readers (called players in this environment) enter the story when they enter the room. The story unfolds as they discover and examine the output and input devices that are contained in objects in the kitchen. The work exists in virtual time and space and not only challenges readers to discover streams of text, but also locates them within the story. To connect to LambdaMOO, telnet to lambda.parc.xerox.com port 8888

8. Many Xerox PARC computer science researchers have Sun SPARC workstations, which are extremely powerful computers, on their desks.

9. Arts Wire is an online communications system for the arts that is sponsored by the New York Foundation for the Arts. It is available at http://www.artswire.org/Artswire/www/awfront.html

10. FORWARD ANYWHERE, the first phase of our work, is in press for Eastgate Systems. Parts of FORWARD ANYWHERE are also available at: http://bush.cs.tamu.edu/~malloy/html/start.html

COMMUNITIES OF INTEREST

Estrogen Brigades and "Big Tits" Threads: Media Fandom Online and Off

Susan Clerc

In 1991, the small college I worked for in western New York went online and gave me an email account. Within a few months, I was up to my armpits in fan fiction and sleeping in hotel rooms with strange women. I'd found media fandom through the net.

It's actually hard to miss fandom on the net, thanks to the dozens of newsgroups and mailing lists devoted to TV series, but the Internet, Usenet, and pay services like America Online and GEnie are only part of the picture, one type of venue for fandom. Off-line fandom existed long before online fandom, and I've found it interesting to explore how they over-lap and diverge and the different roles women play in each.

I have to begin by defining fandom, an intimidating prospect at best since no two fans are likely to agree on anything, much less how they should be described. The task is further complicated in that fandom adds new flavors daily, as the term is applied to anything that forms the basis of a hobby or interest group: sports, music, comic books, even conspiracy theories are all sources for fandoms. The piece of the puzzle I'm involved in is media fandom, an outgrowth of science fiction literature fandom.[1] Now that I've said that, you're probably thinking Trekkies. If you're starting, as I did, with an image of fans as geeky Spock-eared boys who need to "get a life," you're unprepared for the reality of the subculture. Frankly, the rich, complex, untidy reality of fandom bears about as much resemblance to the *Saturday Night Live* stereotype as a robust burgundy bears to cherry Kool-Aid. They're both red liquids that can give you a headache if you overindulge, but there the similarity ends.

One of the first things you have to understand is that fandom embraces a huge number of series and movies, not just the Paramount herd of *Star Trek* cash cows. *Dr. Who, The Professionals, Blake's 7, Sapphire and Steel, Sandbaggers, Robin of Sherwood* (all dead British shows as of this writing), *Babylon 5, The X Files, The Man from U.N.C.L.E., Starsky and Hutch, The Wild, Wild West, Quantum Leap, Wiseguy, Kung Fu: The Legend Continues, Highlander, Forever Knight, Real Ghostbusters, Beauty and the Beast* (the TV series), *VR.5, MacGyver* and *Star Wars* are a few of the many TV shows and movies fans have incorporated into their shared frame of reference. Most fans either hold simultaneous interests in two or more series or progress from one to another in a sort of serial monogamy. All the fans I've met, both online and off, are conversant in several of the current U.S. hits, defunct British programs and cable-resurrected shows.

Fandom is a community, a social network of small groups and individuals scattered across the United States, Canada and other English-speaking countries, created and maintained through overlapping, conflicting, complex ties to each other. Fannish activities are the ties that bind the community together. Though the activities are diverse, many are discursive—the most primal instinct a fan has is to talk to other fans about their common interest.

"Coming Out"

Everyone in the community has what some fans call a "fan coming-out story," a universal tale that inevitably includes a line about how relieved and horrified the fan was when she finally discovered there were others like her out there. My own goes like this: I fell in love with *Blake's 7* in 1987.[2] It wasn't until I got online in 1991, however, that I found out there were other people who knew the show. I couldn't believe it when I saw someone using the name of one of the characters as his *nom du net,* and I was even more amazed to discover from the mailing list SFLovers Digest that people were talking about the series. I honestly thought I was one of very few people who'd seen it and become obsessed by the sparks flying between the title character and his sullen shipmate, Avon. The satisfaction of finding others who wanted to talk about Avon's mental state in the fourth season, whether Blake and Jenna were lovers, and how the women climbed ladders in those heels was tremendous. At the same time I was horrified because I realized how much of fandom I'd already missed and that I could never catch up. Long before I stumbled in through the net, there had been a thriving offline following for *Blake's 7* carried out through a bewildering tangle of print venues that had been born, lived, and often died before I even knew they existed.

Fannish Forums

Before modems made newsgroups, mailing lists and email part of daily life, fans created a number of print venues that let them communicate with each other: APAs (amateur press associations), letterzines and newsletters, all of which differ from each other more in format than in content. APAs, for example, tend to consist of a fixed number of people all of whom supply the editor with enough copies of their contributions (or "tribs") for all the members. The editor then collates and mails the compilation to each individual. A certain level of participation is mandatory for inclusion—no lurking allowed. The result is a fairly close-knit group who feel freer to include personal information than, say, the average poster to a *rec.arts* newsgroup does. Letterzines, on the other hand, work on a subscription basis without a participation requirement, so they're more like net groups in that more people receive and read the letterzine than actively contribute to it. Newsletters, of course, are publications from fan clubs, either local or national. All of these venues, like net groups, include a mix of content, but the amount of irrelevant chatter is much lower in print than online and the number of people involved is usually smaller.

Fans, whether online or off, discuss characterization (if Avon really doesn't like Blake, why does he keep following him?); express their affection for, or dislike of, particular characters; alert each other to appearances by the actors on talk shows or in other roles; compile lists of useless information (for example, versions of Dr. McCoy's "I'm a doctor, not a _____"); speculate about what would have happened if some feature of a story had been different (what if Blake hadn't disappeared after the second season?); compare series, seriously and not so seriously *(Star Trek: The Next Generation* versus the *Love Boat*—bald captain, black bartender, annoying teenage child of crew member); make up drinking games (take one sip every time Agent Mulder from *The X Files* drops his gun); and hash over any number of other issues arising from the aired episodes.

Analyzing the events in specific episodes is probably the most common form of discussion. Fans try to fill in the gaps left by writers and form connections between episodes. An *X Files* letterzine, for instance, might be filled with letters speculating about what happened when Agent Scully was abducted, using evidence from previous and subsequent episodes. In addition to group discussions online and off, fans are compulsive letter-writers, with voluminous one-on-one correspondence. Online, email takes the place of "snail mail." For many fans, net groups are a continuation of off-line practice, the proverbial old wine in a new bottle, because they have always engaged in the sort of long-distance communication we see online.

The social network isn't restricted to remote communication, however: Many fans belong to local groups. Some are formal fan clubs, but many are informal circles of friends who get together on a monthly basis to watch episodes of favorite shows and chat. On a larger scale, many fans also attend conventions ("cons") at least once a year. Cons are a chance for fans to meet each other for the first time, for those who have become acquainted through print or net discussion groups to meet face-to-face, and for friends who live far away from each other to get together.

At my first con—Visions, a British media con held

annually in Chicago—I struck up a conversation with a woman standing next to me in an autograph line; she turned out to be from my area and part of a *Blake's 7* group, and now I commute to Cleveland every month for a party with them. The next con I attended was MediaWest, held every year in Lansing, Michigan; there I roomed with a woman I'd met only online. Now my usual con-going roomie is someone I met online first and in person later at a con (hence my reference to sleeping with strange women in hotel rooms). I look forward to seeing a few long-distance friends at every con.

In media fandom there are essentially two kinds of cons, those with guests (actors, writers and others associated with production of the series) and those without. If guests are present, the convention centers on them, including question and answer sessions, autograph and photo sessions, and often a variety show in which the guests perform. Fan-centered events are also scheduled but often take second place to the celebrities. At cons without guests, all of the events focus on the fans. There are discussion panels, fan artist exhibitions, filking sessions (filk songs are song parodies along the lines of Weird Al Yankovic or Tom Lehrer or sometimes more serious ballads) and costuming contests. Music videos have also become a standard feature at most cons. Fans use two VCRs to edit together clips of one or more series and set the clips to a song. Although similar in definition to professional music videos, the fan videos tend to have a strong narrative structure, using the song to offer a new interpretation of the series and relying on the audience's knowledge of the original context of the clips. Music videos, paintings, costumes and filk are all widespread creative outlets for fannish passion; all of the series mentioned so far, and more, have inspired artists in those fields. However, the best-known and most popular creative outlet is fan fiction.

Fans themselves question why they write and read stories using the characters from TV universes. The most basic answer is "We want *more*." In a sense, fan fiction is an extension of the analysis and speculation fans do in APAs and other print venues. Before I got online, I thought writing fan fiction was as fundamental to anyone with an imagination as eating chocolate is to anyone with a mouth. I'd been doing it all of my life; from the time I was six and my best friend, Peggy, said, "Let's play *Man from U.N.C.L.E.*," I've been rewriting scenes from my favorite shows in my head. It seemed to me the most natural thing in the world to want more of what I liked: more of these characters I loved, putting them in new situations, taking them where the TV writers couldn't or wouldn't, filling in the scenes that were left out. Being a Sherlock Holmes fan encouraged me in this peculiar behavior. Holmes aficionados have generated fan fiction for years: the Apocrypha, stories about Holmes written by famous Holmes fans like Anthony Boucher and Alexander Woolcott, are fan fiction, as are the books about Holmes that continue to appear year after year. All that distinguishes these works (and others like Susan Hill's *Mrs. deWinter*, Alexandra Ripley's *Scarlett* and David Thomson's books *Suspects* and *Silver Light*) from the stories written by fans and privately published within the community is that you can go to the bookstore and buy the titles I've just mentioned.

Amateur fan fiction is technically a violation of copyright, but for decades media fans have published and sold zines (short for "fanzines," which are collections of stories) at cons and through the mail without interference from corporate copyright holders. Fan fiction is not for profit; zines are sold for production cost, which is the legal loophole that supposedly guards fan fiction writers from prosecution, though no one knows for sure because there's never been a court case.

What Are All These Women Doing Here?

Almost all fan fiction is written by women, which leads to another very important point about off-line fandom: The majority of media fans are women. Women write and read almost all of the fan fiction, make the music videos, create the artwork, organize and attend conventions, run APAs and letterzines and belong to fan groups—they are actively involved, in greater numbers than men, in every facet of media fandom. Media fandom wouldn't exist without women because more women than men do the communication work necessary to forge and sustain the community. The public impression that males dominate fan activities is largely the result of outsiders' emphasis on *Star Trek* fandom, which does seem to consist of more males than females. But this emphasis misses the nature of the fannish subculture as a whole. The misconception that males dominate media fandom is also online-fostered; there are simply more men than women online. My composite friend, Mary Sue, illustrates why women play a much less prominent role in online fandom than they do in fandom off-line.

Getting Wired

Mary Sue has been active in off-line *Robin of Sherwood* (*RoS*) fandom for several years. She writes fan fiction, belongs to an APA, attends conventions at least twice a year and corresponds with a dozen fan friends. She regularly gets together for a bash with local fans to talk and watch episodes of *RoS*. Like many other series that fans have taken a shine to, *RoS* lives on in fandom thanks to the extensive network of video-tape sharing within the larger fan community. Tape sharing is the way fans pimp for their favorite series, especially the ones that are no longer on the air: They send friends they want to lure into fandom episodes they hope will hit their kink ("this one guy is such a bitch, you'll love it," "it has story arcs and the characters really develop over time," "the relationship between the two male leads is intense").[5] The more people hooked on the series, the greater the chance that stories and art will be generated. And of course, the more people to discuss the show with, the better.

To pimp for her current favorite series and to create music videos and gather information for discussion or fan fiction writing, Mary Sue has two VCRs and a TV. The speed with which she can connect, dismantle, rearrange and reconnect the equipment would put an Indy 500 pit crew to shame. These machines are a vital part of her fannish life. She also has a computer she uses to write fan fiction and letters to her numerous fan friends, and perhaps even to produce a fanzine. Like the TV and VCRs, the home computer is a tool that lets her participate fully in the community. What Mary Sue didn't have when I met her was a connection to the Internet. But she didn't think she needed one: Of her dozens of acquaintances, only a handful were online and they were still reachable by snail mail and telephone.

Like most of Mary Sue's friends with net access, I got my account through work and it was free (I just had to avoid my snooping employers). We all raved about how fast email was compared to regular mail, how lively the big mailing lists and newsgroups were, how easily pictures and sound files and transcripts of episodes and other goodies were available through FTP, and we all urged Mary Sue to get a modem and join America Online, GEnie, CompuServe or one of the other commercial services, although this would take a little money to begin with. "What money?" she demurred.

A year went by, and more and more of Mary Sue's friends went online and it began to seem like a good move to her, too; the advantages of being able to communicate quickly and easily with a large number of friends were starting to outweigh the costs and headaches of getting set up. During that same time, her husband tinkered with their computer and installed a modem. When Mary Sue asked how it worked, though, he didn't have time to help her. Like many fan women in her position, Mary Sue finally turned to friends who were already wired for help. The informal support network fandom provides for its members came to the rescue as friends offered advice about freenets, online editors and subscribing to mailing lists, and even camped over for weekend tutorials on using the new medium.

The point of this scenario is that fan women, although mechanically proficient and technologically savvy compared to the mainstream population, suffer from the same societal attitudes about gender and technology as everyone else. If a gadget serves a purpose, especially one that furthers the traditional feminine role of creating and maintaining community, women will take to it readily. Unlike men, however, they are not inclined to tinker with a new machine for the sake of tinkering. Women are also at an economic disadvantage: With less disposable income, they are not as likely as men to experiment with modems and software they aren't familiar with. Fan women may have an additional deterrent in that they're already extremely well-connected off-line to a large number of other fans. For them, there is little benefit to net access unless many of their friends have it. When that critical mass is reached and it becomes beneficial to go online, fan women will likely turn to other female fans as an informal support network who can give in-person tutorials rather than to distant male technicians.

Most fan women enter the net through work, because a male relative has set up the equipment at home or because their friends have access and encourage and help them. Their delayed entry (compared to males) into cyberspace is reflected in statistics about time online and age. Women who answered a survey I posted to several fannish groups in 1993 about fan net use[4] had been online for less time than the men who answered: Thirty-seven percent of the women had been online less than a year, compared to 24 percent of the men. The disparity was particularly noticeable among people who had had access for more than four years: Only 14 percent of the women fell into this group, compared to 29 percent of the men. The women also tended to be older, suggesting that they were already out in the labor market when they gained access: most were 23–30 or 31–40 (37 percent and 35 percent), while most of the men fell into the 18–22 and 23–30 ranges (38 percent and 39.5 percent); the percentage of women over 40 is double that of men in the same age group (15.8 percent to 7.9 percent).

Going Public

Once they're online, fan women participate in public mailing lists and newsgroups less than you'd expect for such a communicative bunch, although women who have been active in fandom off-line may have an edge over those coming into the community through the net. Those who are already fans have a tradition of female participation behind them and are likely to find familiar names waiting for them online, factors that may increase their confidence about posting. Yet they still post to public groups less often than men do. Women, regardless of their previous fannish experience, just don't talk as much in public as men, partly because they are socialized against drawing attention to themselves by seizing and holding center stage.

If a newly online fan does pop her cherry and post, reaction to her maiden messages will affect her future posting rate. If everyone ignores the post, the newcomer may interpret the lack of response as a sign she's not welcome. However, not responding shouldn't necessarily be construed as a conscious attempt to bar women from discussion; it's often, rather, a sign of reluctance by members of both sexes to waste time on people who may not be in for the long haul. But getting no response does tend to happen more often to women than men. Susan Herring's research on gender-linked performance on academic mailing lists, which demonstrated that both men and women tend to respond more often to men than to women, even when the topic is more relevant to women, indicates that lingering social mores about women affect response.[5]

Each person has her or his own reason for not responding to a poster's first efforts, but for the poster languishing in the silence it feels like rejection. If a poster complains that she's never responded to, list members may be more conscientious about offering back chat for a few weeks, but this effort eventually tapers off because no one really wants to post or read an endless stream of *me, too*'s. A complaint may generate private email rather than spark list discussion, in which case the intended goal of list conversation isn't met, but the list member is still made welcome. Once she has that "off-stage" backing, she may participate more.

A direct ploy for dissuading women from posting is to send them offensive email. Women's magazines and other popular media have widely disseminated the idea that women are routinely hit on or subjected to crude insults on the net. These exaggerated accounts may deter women from initially posting as much, as flame mail may silence them once they've spoken up. I've never received a "wanna fuck" (a name I've seen used on one of the *alt.sex* newsgroups for this sort of obscene mail) in the time I've been online, although I've participated "loudly" in several groups. From what others have told me, offensive email to women happens more often on the pay services than on the Internet and Usenet.

What does happen with disheartening frequency on Usenet is the "big tits" thread. Depending on the specific newsgroup, readers are assaulted from time to time by posts from sexually and emotionally immature boys about Scully's big tits, Deanna's big tits, Beverly's big tits, Peri's big tits, Janeway's lack of big tits and so on. Many women are put off by the obsession boys online seem to have with actresses' breasts and their compulsion for discussing them in public. Fortunately, big tits are not the dominant matter of discussion on any of the fan-related boards. In fact, the subject is often easy to miss among the insightful, perceptive posts from both men and women about character motivations and what might happen in upcoming episodes. Unfortunately, a large number of juvenile "big tits" posts can make women feel extremely unwelcome and/or threatened, especially when the message is hostile and moves from mammaphilia to graphic descriptions of what the poster wants to do to the actress or character. Women themselves drool over the actors and characters, of course, but their posts are rarely of the "I want to assfuck Ro Laren" variety fanboys like to post.

As annoying as "big tits" (and the like) posts are, it's essential to remember that they, and the problem of being ignored after posting your deepest thoughts, are minor worries—speed bumps rather than barrier walls. Many, many women do post actively on newsgroups and lists like *alt.tv.x-files, rec.arts.sf.tv.quantum-leap*, the *Blake's 7* list and STREK-L (the large Internet *Star Trek* list). There are dozens of newsgroups and mailing lists out there, and women are on all of them, although women favor mailing lists while men go for the high-profile Usenet newsgroups. According to the survey I conducted, 42 percent of the women posted at least once a week to public lists, but only 22 percent of the men did. In comparison, the weekly posting rates to newsgroups were 28 percent for women and 37 percent for men.

The difference is clearly the format and not the series: For example, although women are a very strong presence on STREK-L, they are less than a third of the posters on the newsgroups *rec.arts.startrek.current* and *rec.arts.startrek.misc*. This is also true on the *Dr. Who* newsgroup and list. The *Babylon 5* mailing list has a reasonably high percentage of women participants, but the newsgroup was almost exclusively male until very recently; and even now only a handful of women post regularly compared to the literally scores of men. The *Red Dwarf* newsgroup is almost completely male although there are certainly women fans of the series. The only newsgroup where women seem to account for half or more of posts is *rec.arts.sf.tv.quantum-leap*, and that series doesn't have a public mailing list. The other fannish newsgroups range in between but always have fewer women than men. On several mailing lists, however, women are the majority of posters or a very vocal minority: The *Blake's 7* and *X Files* mailing lists seem to have more female than male contributors, and the *Star Trek, Highlander* and *Dr. Who* lists have a plethora of outspoken women contributors. Some series that attract primarily female fans do not have online discussion groups: Neither *Beauty and the Beast* (the TV series) nor *The Professionals*, two series more popular with women than men, has a public list or group devoted to it even though both generate a lot of fan fiction and discussion off-line.[6]

It's interesting to note that the last three times the question of creating a newsgroup was raised on the *Blake's 7* mailing list, it was raised by a male and contested primarily by women, which indicates a tension between male and female goals in communication. To grossly generalize, men communicate for status, and women communicate to maintain relationships.[7] This was spontaneously confirmed when several women responding to my survey cited a sense of community as one of the most valuable aspects of net groups:

I would not say my interests as a fan have grown quantitatively, but instead have changed qualitatively. I do not love B7 [*Blake's 7*], for example anymore than I did before, but I now feel less like an individual fan than a member of a community. When I talk about the *Star Trek* movies, for example, I tend to say things like "most of *us* hated the fifth one" or "*we* grieved when Gene died."...I like feeling there are many kindred spirits out there who are supportive of my interests....I like feeling very connected to all of the other fans out there.

Although some newsgroups manage to attain a sense of community, mailing lists are more likely to do so because of the way they're set up: Fewer people post to them, so there's a feeling of familiarity (sometimes you wanna go where everybody knows your name), messages come directly to the subscriber as email, listowners maintain subscriber lists, so there's a record of who lurks as well as who posts, and many lists have archives that let subscribers access list history. In contrast, newsgroups have more posters and more anonymity—some posters use anonymous posting services, several online reading systems leave off the posters' names, browsers don't leave a trace, and posts disappear after a few days. Perhaps more importantly, as a couple of fan friends recently pointed out to me, you have to come out of the fan closet to join a mailing list: You can't pretend you're only casually interested in *The X Files* when there are fifty messages about it in your mailbox every morning. Commitment to the group, however mild, plus the personal admission, however casual, that the series does mean something to you, contributes to the lower noise level on lists. There are fewer "trolls" (people who post chain letters, "big tits" posts and the like, solely to get a rise out of people). Many people, like the fan quoted below, perceive a qualitative difference between conversation in the two formats:

To be quite frank I have found the mailing list discussions I have had re: the series to be far more intelligent than *anything* I have read in the newsgroup.

Whether they choose newsgroups or mailing lists as their main net neighborhood, women who post may run into an unexpected response: Off-line, the presence of women in media fandom for any series is taken for granted; online, where there are large numbers of males unfamiliar with this tradition, the presence of women raises virtual eyebrows. Any public speech by women seems to stand out and often leads to the perception that women are talking more than men when, statistically, they aren't.[8] When women appear to be in the majority online, someone is bound to ask "What are all these women doing here?" The question is usually accompanied by a statement like "I thought this series appealed mostly to guys, but most of the posts so far have been from the fair sex." During one week, in fact, versions of this question appeared simultaneously on both the *X Files* and *Dr. Who* mailing lists, and it appeared on the *Highlander* newsgroup while this essay was being written. Regardless of the poster's intent, the message behind the message is that "this show is a guy-thing; there's something wrong with you for liking it." The question may intimidate some women but usually leads to a barrage of posts along the lines of "What is it you think women don't like: well-written scripts, witty dialogue, great characters?" and other rebuttals.

Going Private

When gender differences are apparent on public lists and newsgroups, they tend to occur along traditional lines: Women focus on character issues while men more often discuss hardware and special effects. When divisions occur, and this is important to keep in mind, both sides usually continue to co-exist without rancor. Sparks can fly, however, when sex becomes an issue, as when women lust after actors or lament the dearth of good female roles in the series.

When women begin sharing lustful thoughts about the actors, the response is somewhat hypocritical. A common, if mild, masculine response is "How would you ladies like it if us guys started slobbering on about the actresses?"—a remark that blatantly ignores the rampant slobbering in the perennial "big tits" threads. It also overlooks the fact that many of the posts about actors come from women who have contributed a lot to the group and who usually post more substantial messages, and that their comments are considerably more friendly toward the lust object than the men's posts tend to be.

More controversy is caused by comments about the lack of decent roles for women in TV series. After a few months of posting to STREK-L, I complained that all of the women in the series, supposedly in the distant future, use their husbands' last names (someone recently posted a similar objection on the *Babylon 5* newsgroup). I also commented that the roles given to women were still in the nurturing ghetto (doctors, therapists, bartenders) and that the only really forceful women were either evil or killed off, or both. Many women have made the same observations before and since in the *Star Trek* groups, and the typical response is a lengthy diatribe whining about white males being oppressed by politically correct minorities and declaring that politics should be left out of the discussion. The same objections, though often more rancorous, greet many discussions of homosexuality (gay fans have long called for representation in the *Star Trek* universe). *Star Trek*, say the heterosexual white boys, is too politically correct. Their naiveté would be touching if it weren't so depressing.

Rather than giving up on public groups entirely in the wake of such responses, many women have formed private mailing lists that combine the intimacy of small groups with a support network similar to the kind fan women create offline. One of the first and best-known examples of the private lists is the satirically named Star Fleet Ladies Auxiliary and Embroidery/Baking Society, started by Janis Cortese after she expressed dissatisfaction with the women's roles on *Star Trek: The Next Generation* and was flamed for it.[9] The SFLA&E/BS

carries on its own conversations independently of the large groups, but its members also participate in those groups and their reviews of episodes are always entertaining and thought-provoking. Their review of the *Deep Space 9* episode "Through the Looking Glass," for example, had this to say about the portrayal of Kira in a mirror universe:

People who are queer aren't queer to be weird or deviant or evil, and WOMEN who are queer aren't queer to make pre- and post-adolescent boys' palms get sweaty. This was pathetic—cartoon "Penthouse" lesbo-a-go-go crapola. Paramount, you may think you did something "daring" or whatever with this, but trust us on this one—ooh-ooh-squeal scenes in which lesbian or bisexual women are presented as unstable, dangerous and oversexed are NOT AT ALL DARING *OR* FORWARD-THINKING. In fact, it is just about as reactionary and BORING as you can get.

Many of us regret that the SFLA&E/BS never tell us what they really think.

The SFLA&E/BS is one of a growing number of private lists sprouting from larger groups. There are also several self-named "estrogen brigades," small private mailing lists where women gather to discuss a favorite actor and other topics. Like Cortese's group, the estrogen brigade members post both within their own mailing list and to the big mixed-sex groups that spawned them. A brigade's name, attached to the posts to the larger groups alerts other women to the existence of alternative discussion, and acts as a defiant commentary on the "nice girls don't act that way" double standard. The Patrick Stewart Estrogen Brigade (PSEB) seems to have been the first, but there are others; two I've seen frequently in .sig files are the David Duchovny Estrogen Brigade (DDEB) and the Siddig el Fadil Estrogen Brigade (SEFEB). (.Sig files are signature files, lines some people append to their posts that contain their address and often a quote or other clue about their personality and philosophy of life.)

Conversation within these groups is by no means limited to drooling over actors, however. As one fan said:

the list I'm on is the Duchovniks, which makes *no* attempt to hide the fact that we started primarily as a place to ooh and ahh over the magnetism of DD [David Duchovny]. We are NOT ruled by our hormones—they merely brought us together as a group.

We have since found far more to bring us together than our interest in DD.

One of the most valued results of smaller women-only (or women-dominated) groups is the intimacy that develops between the members. As one of the members of the SFLA&E/BS said: "I'm part of one of the longest-running slumber parties in history."[10]

To some degree, the small online groups are computerized versions of traditional off-line fandom forms, like APAs, through which fan women have always found each other. They use a new medium to duplicate the webs of friendship that make up the heart of fandom. Since fan women online are a mix of women who are already fans off-line and those just entering the community through the net, it isn't clear whether these smaller online groups spring from traditional fan customs or from a more general feminine desire for community and intimacy, and it may be impossible to distinguish between those sources. Either way, the private lists created by fan women show their resourcefulness and creativity in adapting the medium to their own needs.

Let's Get Physical

Fun and rewarding as communicating via the Internet can be, many women want to extend virtual friendships into real life. The overwhelming majority of fans who answered my survey, male and female, exchanged private email with seven or more people. Women, however, were likelier to meet email friends in person (50 percent of women and 41 percent of men had met some of the correspondents).

Meeting in person can be problematic; according to one woman whose nonfan group arranged a meeting, "[w]hen you get these compressed messages, you fill in the blanks and think you've met your soulmate, but in fact a lot of these people have trouble functioning normally."[11] Horror stories among media fans are rare, however, perhaps because fandom offers a unique venue that helps smooth the crossover from virtual to real-life meetings: conventions. Conventions are large, public gatherings that all concerned were going to attend anyway. Meeting each other is a bonus, a short-term event in the midst of a larger event of interest, not a stressful and high-pressure platonic version of a blind date.

Another factor in the success of fan meetings is the rich tradition of long-distance fan friendships created and carried out through APAs, letterzines, newsletters and private correspondence. As a result, fans usually have happy memories of meeting online acquaintances.

I coerced my friend Joan into attending her first con because she would get to meet people we'd both been talking to online for more than a year (that is, I wanted gossip about them). I was also eager for Joan to attend because my own experiences meeting people in person at cons had turned out surprisingly well. When I went to ZCon in 1993, I didn't know any of the women I'd agreed to share a room with except as names on a computer screen, and I had a marvelous time. It was awkward at first because even though you do know someone fairly well once you've exchanged slave fantasies with them online, you still have the shyness of any other first-time meeting and have to adjust to how the person looks and talks ...and snores. When my friend returned from the con, she summed up the experience:

I felt like these were women I had known since childhood. Because of the list , we had a head start on each other's personalities. It was so odd to be with people I didn't know and yet I did....By the time I left Sunday (an hour late because I hated to go), I was so glad I had gone. There had been 16 list members there ...and I liked all but one.... So I'm packed for Visions [another convention] ... when do we leave?

Fan Fiction

The majority of fans online participate in at least one fan activity aside from net groups (82 percent of women, 57 percent of men) and many of them started fan activities before getting online. Con-going is the most popular for old-timers and newbies alike. The second most popular—and the one that brings men and women fans into conflict with each other and fandom into conflict with the outside world—is fan fiction.

Fan fiction online has mushroomed in the last few years. There are at least four newsgroups dedicated to story posting, and at least three mailing lists. In addition, stories occasionally appear on discussion-oriented mailing lists when members decide to do a round robin (list members take turns writing sections of a story), and FTP sites around the world archive stories from lists and groups. Anonymous FTP sites allow people who can't access the groups to recover the stories, as well as pictures, sounds and other goodies (some archives have transcripts of episodes and lists of fanzine editors, for instance).

Unlike the vast majority of print fan fiction, a lot of online stories are written by young men, many of whom have no knowledge of the off-line community and the history of fan fiction written by women. Opinions of the quality of their online fan fiction tend to be low among women with experience in fandom:

I looked at some of the ST:TOS stuff in the archive and online and hated it. I thought it was dire and also boring, since most of it wasn't about K[irk] and S[pock], but about random characters created by boys who really just wanted to be ST [StarTrek] characters themselves

The reasons for distaste are essentially twofold. One is referred to in the post above: the tendency of men to include themselves in their stories. Women who write fan fiction have long held this practice in disdain, sometimes criticizing any strong original female character as being nothing more than a stand-in for the author. Telling stories about themselves seems to be part of a male aesthetic, though, since it often happens in the male-dominated field of minicomics as well and is reflected in story-telling practices among men in general—when asked to tell a story, men talk about themselves and women talk about other people.[12]

The other characteristic about "Boys' Own Stories" that turns off many female readers is an excessive interest in hardware, violence, and convoluted plots that go nowhere. Fan fiction written by and for women has always focused primarily on the characters' relationships. Of course plots are important, but they are used to explore the nature of the characters (what would Spock do in this situation?) not just for the sake of creating something cool to blow up. I'm exaggerating here, of course. Yes, many men do write stories that deal sensitively and perceptively with the characters, and many women write stories about destroying horribly beweaponed aliens; but with limited time online to choose what to read, many fans fall back on the reality that a great deal of fan fiction can be divided along gender lines:

I am a fan off-line, too, and I am a big *X Files* fan now. But when I cruise The X Files .creative list, which has a pretty high volume of original stories, I just ignore the fanfic by male posters. I never expect the boys to have the same interest or fan aesthetic that we have, since in my experience of *Blake's 7* fanfiction (the paper variety, not the digital) the best stories were by women. I usually have limited time to cruise Usenet anyway, and I want to maximize my chance of hitting something of value, so women's stories are the best shot I have.

The popularity and high visibility of fan fiction online have caused problems between fans and series-related people who are also all over the net these days: producers of the series fans cluster around and corporate copyright holders.

One of the things many fans, especially men, like about net fandom is that writers, actors and others involved in the series are also online. It's exciting to have discussions with someone who's responsible for creating a TV series, to hear his opinions and chum around with him. On the other hand, it can also have a chilling effect on the basic nature of fan debate—discussing interpretations of the series. J. Michael Straczynski (aka Joe or JMS), the creator of *Babylon 5*, is a regular participant on *rec.arts.sf.tv.babylon5*. The condition for his staying on the newsgroup is that no story ideas appear there because it would open him up for lawsuits and the lawyers for PTEN (Prime Time Entertainment Network, the pockets that fund the series) would force him to pull out. To most fans, his presence is well worth the restriction:

This is truly a unique situation we have here and I am VERY glad to have the opportunity to discuss the show with the Great Maker. You must admit that everyone who reads this newsgroup has a much greater understanding of and appreciation for the show primarily because of the insights and answers provided us by Joe. Woe to the person/persons who destroy that line of communication by posting fanfic here!

To others, though, it's an unwelcome limit on their options. Fan fiction is all about looking at a series from different angles and playing with possibilities left open on screen; if questions that could fuel this sort of fan play can be resolved with finality by deferring to the series creator, the game is ruined. The ban on fan fiction also interferes with one of the basic activities of fandom—wondering what will happen next. As one poster pointed out, many of the most interesting threads on the newsgroup arise from the sort of conversation that can easily cross over into the forbidden realm of story ideas:

Given how long it takes for the story we are watching to get anywhere, speculation about future directions of the show is—in case no one has noticed—one of the meatier topics of discussion around here. "What do the Shadows want?" "What do the Vorlon look like?" "What will the new Minbari government do?" "What is Bureau 13 up to?" "What are the extent of Talia's powers?" [..] "The real reason the Minbari surrendered is that they wanted our hair care secrets and were afraid nuclear bombardment would destroy them forever."

Aside from the PTEN/JMS ban, Paramount demanded that America Online and Prodigy pull *Star Trek* fan fiction from its databases and prohibit their customers from posting any more. But since there are still several Usenet newsgroups, Internet mailing lists (including one for *Babylon 5* fan fiction), private mailing lists that distribute fan fiction, and publicly accessible FTP sites where fans can find stories, it's tempting to think outside forces have no influence on fans at all.

But that isn't true. There's a depressing amount of sucking up to JMS, for example. Just the fact that the "don't post story ideas or Joe will leave" thread seems to be going on in perpetuity shows that his presence means a lot to many of the people on the newsgroup. If that weren't enough, the kiss-up "great flame, JMS" posts following some of his messages and the huge number of posts flagged "ATTN: JMS" should clue any reader to the fact that he wields a lot of influence.[13]

The direct access that the net provides producers and fans to each other can have positive aspects as well. By lurking anonymously on lists and newsgroups, those involved in the creation of series can get immediate, unmediated reactions to their work. They can also use the medium to rally the troops quickly and efficiently if a series is in danger of cancellation or has already been canceled but has a chance of resurrection in syndication.

For fans, the unprecedented access to The Powers That Be on their favorite series tends to create a sense of connection and participation with the series. Just knowing that a key writer or other behind-the-scenes figure is aware of fans online is enough to create the strong feeling that one's voice might be heard. That's a potent brew for most fans and it's nothing compared to the good vibrations that rock the net after someone behind the scenes acknowledges online fans by mentioning the net in an off-line interview or dropping into an online chat on a pay service. While this much recognition is becoming fairly routine as the net spreads to more and more of the population, the love affair between X Philes (as *X Files* fans online call themselves) and the series production team is still rare. The series has at least twice given an open nod to its online fans: In the second season premiere the names of several X Philes appeared on a passenger list and an early third season episode was dedicated to a fan who had run a discussion group on America Online.

On the other hand, the sheer volume of fan talk and its high visibility on the net has led to some misunderstandings and hurt feelings. In a magazine interview, David Duchovny expressed somewhat negative feelings towards the activities of female fans online. When the remarks were relayed on the newsgroup, a long conversation ensued about the relationship between actors and fans and how appreciation should be expressed. Although some fans decried drool, many supported salivation:

DDEBers [David Duchovny Estrogen Brigade] are reeling in shock from the news that DD apparently "hates" them. Now, I missed the first Speedo-go-round, but I lived through the second (as well as the "what is DD's best feature/quality?" thread). It sounds like good, clean (well, maybe a tad racy) fun. Still, we've seen numerous apologies to DD, CC [Chris Carter, creator of *X-Files*], et al, about it. But hey, what exactly is wrong with noticing that he's a handsome man? He works out, and it shows. [....] Someone else on the conference rightly pointed out that the man is probably just slightly embarrassed by the infatuation: and this is GOOD—humility is far more attractive than vanity. The long and short of it >:) is, there is no need for you ladies (and gentlemen?) to fall on your swords just yet.

My first reaction to this controversy was "WOMEN? Lust-ing after an ACTOR? Who'd'a thunk it?" Actors are sex symbols and their livelihoods do rather rely on their physical attributes. Think of the career Ernest Borgnine might have had if that weren't true. At the same time, like most fans I can sympathize with an actor's dismay, and perhaps fear, at being confronted with large-scale drooling over his appearance in a bathing suit. All the more reason, in my opinion, for actors and others who aren't in the community to restrict their interaction with fans. They can go to conventions or the pay service chat sessions where it's safe and controlled, but shouldn't wander into fan groups and imagine scrolling one screen gives them a real idea of the community. And fans should understand that the actors' interests as professionals doing a job and our interests as fans are often at odds; they shouldn't be expected to approve or even understand the ways we play with the toys they've given us.

Slash—Could the Frontier Be Closed?

The issues surrounding online fan fiction, the differences between male and female fans' interests, and the phenomenon of women expressing overtly lustful thoughts all come together in slash, the genre of fan fiction based on homoerotic relationships between male characters. Slash,[14] even more than the other kinds of fan fiction, is written almost exclusively by and for women. It isn't gay porn, although there are some similarities; it's written with an eye to feminine sensibilities—lots of touching and talking along with the fucking—and sometimes it's so inexplicit you wouldn't know it was slash except for the label.

Slash discussions on public groups, like discussions of homosexuality in a series universe, have been known to lead to flame wars. They don't have to: People are quite capable of having rational discussions in which everyone disagrees on everything except their right to disagree, and all of the combatants remain civil. The first time slash surfaced as a topic on the *Blake's 7* list, in fact, we had a great time talking about why we liked or disliked it. The next three times, the conversations became decidedly more heated and less informative. The arguments for and against became so repetitive that one of the women on the list devised a General Slash Defense Form Letter listing all of the oft-recycled objections with the best of the rebuttals. It's kept in the list archives with plans to send it to the next person who expresses his outrage that "such a disgusting thing as slash" exists. Notice I said *his* outrage. As a slash fan, I'd like to categorize all anti-slashers as sexually insecure adolescent males, but that isn't the case. There are women who don't like slash, too. Nevertheless, the worst of the anti-slash posts and the highest level of intolerance do seem to come from young males.

Other fans are not the only problem for slash fans online. We also stand to suffer disproportionately from the wrath of crusading conservatives who think the Internet is the twentieth-century's Sodom and Gomorrah. Some stories have appeared on *alt.sex.startrek.fetish* and *alt.startrek.creative*, but a lot of slash activity is conducted through private mailing lists for story distribution or chat. We aren't ashamed of our fondness for two men in passionate clinches, but many of us are concerned that online "decency" bills will be used against us and the kinds of adult conversations we're interested in rather than against the juvenile drivel that makes some newsgroups a waste of time.

It seems quite realistic to fear that serious talks about the sexuality of Odo from *Deep Space 9*, for instance, will be threatened while the "big tits" threads will be left alone, since anti-pornography laws are often turned against gay and lesbian erotica rather than the violent and degrading heterosexual male material they were originally intended to regulate. More frightening than censorship of the public groups are proposals by lawmakers which would allow for the scrutiny of private email by "Big Brother"—who, we imagine, is not likely to be as turned on by stories about Garak and Bashir on *Deep Space 9* falling in love as we are. But if this ever happens, slash fans will find ways to circumvent it, just as fan women have circumvented copyright, unhelpful computer services flunkies, and nasty little boys, to keep doing what we've always done—talk to each other about the series we love.

At the beginning of this essay I said that net fandom was only part of fandom. One of the key differences between on- and off-line fandom is the role women play in each. Attitudes toward technology and communication have delayed some women's access to the net and prevented others from participating as actively online as they do off-line, with the result that women fans are not as prominent on the net as they are in real-life fandom. But, women fans have dealt with the disadvantages creatively and ingeniously, adapting the new medium to their own needs as well as adapting off-line fannish customs to the net. Whatever roadblocks are thrown across the Information Superhighway in the future, I know the women of fandom will find a way to overcome them if not slash right through them.

Endnotes

1. For more information on media fandom, see Henry Jenkins, *Textual Poachers: Television Fans and Participatory Culture* (New York: Routledge, 1992) and Camille Bacon-Smith, *Enterprising Women: Television Fandom and the Creation of Popular Myth* (Philadelphia: University of Pennsylvania Press, 1992).

2. *Blake's 7* is a British science fiction series that aired on the BBC 1978-1981. Created by Terry Nation of *Dr. Who* Dalek fame (overgrown salt shakers that shout "Exterminate, exterminate!") the series is characterized by moral ambiguity, bad special effects, and conflict between the nominal good guys (all but one are convicted felons)—especially Blake, a rebel intent on destroying the evil totalitarian government, and Avon, a cynical computer expert. During the series several of the regular characters are killed off or depart in unexplained circumstances, including Blake. Continuing story lines allowed the characters to develop more realistically than in most contemporary U.S. action adventure series, where last week's revelation of a tragic past no longer affects this week's hero. The pessimistic feel of the show (nice guys don't just finish last, they aren't even very nice) is a refreshing change from compulsory happy endings and quick resolutions. Before *X Files* made the line famous, *Blake's 7* created a universe where "trust no one" wasn't just good advice, it was a way of life.

3. Some series have followings only because fans incorrigibly tape and share; *Blake's 7* had fans in the United States before it ever popped up on PBS because fans in the UK sent camera copies to friends and family in the States. To get around conversion problems created by incompatible broadcast systems in the two countries, people would set up NTSC camcorders in front of PAL TV sets while the episode was on. *RoS* aired in the United States several times but only on local PBS stations and a cable network, so access was limited; tape sharing made the fandom possible.

4. The survey was originally done for my thesis: "The Influence of Computer-Mediated Communication on Science Fiction Media

Fandom," Master's thesis, Bowling Green State University, 1994.

5. Susan Herring, "Gender and Democracy in Computer-Mediated Communication," *Electronic Journal of Communication/La Revue Electronique de Communication* 3.2 (1993).

6. Women may shy away from creating public groups for themselves because they rarely have control of a large list or newsgroup; another reason may be their reluctance to approach their system administrator to request help or permission, assuming, often correctly, that the request will be dismissed as trivial.

7. Deborah Tannen, *You Just Don't Understand* (New York: Ballantine, 1990), 77 and elsewhere. One of the themes of the book is that men's need for independence and women's need for intimacy contribute to different styles of communication.

8. Ibid., 77 (citing Dale Spender).

9. Barbara Kantrowitz, "Men, Women and Computers," *Newsweek*, 16 May 1994, 48–55.

10. Ibid, 50.

11. William Grimes, "Computer as Cultural Tool: Chatter Mounts on Every Topic," *New York Times,* 1 Dec. 1992, C13.

12. Tannen, *You Just Don't Understand*, 177.

13. Since this essay was originally written, JMS has removed himself from the *Babylon 5* Usenet group. Ironically, given the fears of many posters, his withdrawal from active participation in the group had nothing to do with fan fiction.

14. So named because of the fan convention of writing the names of the characters to be paired: K/S is Kirk and Spock as lovers, for example.

Elites, Lamers, Narcs and Whores:
Exploring the Computer Underground

Netta "grayarea" Gilboa

Finding the computer underground is a snap if you know where to look and almost impossible if you don't. Security experts estimate that 35,000 active hackers are out there, with roughly 10 percent more joining the scene per year. The exact number is hard to determine as many hackers work alone, never making themselves visible. There are also regular, organized meetings of hackers[1] and phone phreaks (telephone wizards who manipulate phones and phone companies for their own gain) sponsored by *2600* magazine on the first Friday of every month in many large American cities, as well as in countries like Argentina, England, Germany, Spain and Sweden. Typically the meetings take place in malls, train stations and other public places, with ten to seventy people attending, mostly white males between the ages of twelve and twenty-eight. Numerous weekend conventions, "Cons," have speakers (including members of law enforcement), hungry media eager to sensationalize the underground, and legendary, shadowy hackers who surface for the party. Attendance at major cons–HoHoCon, PumpCon, SummerCon and DefCon–ranges from fifty-five to thirteen hundred people.

Hackers also communicate with each other by modem through bulletin board systems (the largest three of these are Unphamiliar Territory, SecTec and Planet X) and many organized, private channels such as #hack, #phreak, #warez, #virus, #cellular, etc. on Internet Relay Chat (IRC), the anonymous chat facility available across the net. There are also private FTP (File Transfer Protocol) and FSP sites[2] where items traded include password files to major sites, source code, scripts used to obtain root privileges on UNIX machines or to annoy people on IRC,[3] stolen personal email, and pirated software (including software not yet released to the public). A newer trend among hackers is creating private mailing lists to discuss specific technical subjects like altering cellular phones. These mailing lists tend to come and go quickly, since, with all the hackers monitoring each other, nothing said on the net stays private for very long.

Hackers also talk to each other "voice" through illegal teleconference calls that have fifteen or so hackers and phreakers on at a time, on regular or three-way calls often made with toll fraud devices known as "red boxes" (a gadget that simulates the sound of quarters being fed into a pay phone, used to avoid paying for phone calls), "blue boxes" and "beige boxes," or by accessing someone else's PBX system or calling card number. One learns about these phone devices primarily from well-known publications like *Phrack* (an electronic magazine) and *2600* (a paper magazine known as "The Hacker Quarterly"); there are also hundreds of other electronic magazines (most of which do not last beyond a few issues), other paper magazines like *Blacklisted! 411, Iron Feather Journal*, photocopied copies of *TAP* (now out of print) and short information-filled text files, which can be downloaded from BBSs or found on FTP and World Wide Web sites. The determined hacker must be well-read, as tips and bugs may be found anywhere from the latest post in a security newsgroup to a decade-old text file originally written for a forty-column Atari computer.

Newcomers join the scene in a number of ways. They

may stumble across the underground by accident (such as by buying a copy of *2600* and attending a meeting) or be introduced by someone already involved. Most hackers start by using a modem and calling a BBS that has pirated software on it. They either then meet with other hackers in person or get invited to call somewhere else. Newcomers usually don't know to hide their real data, and once they've told a few people, or even just one, their real name, phone number and address, they will have gotten themselves added to hundreds of hackers' databases as well as to the files and databases of various three-letter law enforcement agencies.

Once you're in the scene, it's almost impossible to leave it, because when you communicate "voice" with other hackers they usually tape your calls, and if you have attended a con your photograph is probably circulating out there, too. (Attending a con implies you give permission to have your photograph taken for the con organizer's videotape, and it really is a game to see who ends up on film caught doing what.) Some hackers are public figures and don't mind having their photograph spread across the net. Other hackers live in fear of this happening.

Newcomers are often ignored, harassed, insulted, or mistrusted. They may be tolerated and assisted, but only up to a point. You can get very far very fast, though, if you come to the scene with things to trade (like an endless supply of calling card numbers or a new script that works), publish an ezine (it is a status symbol to publish electronically, but you are considered threatening if you publish about hackers in print) or if you are introduced to the scene by a well-respected member of the community as I was.

My Introduction to the Underground

My introduction to the computer underground was really accidental. If I'd had any idea what I was walking into, I would never have ventured in, or I would have lied about who I was and what my intentions were. God knows the people I befriended sure lied to me.

My first contact was in the early 1970s. I saw calling card numbers printed in a newspaper put out by the Yippies (a political group founded by Abbie Hoffman) and attempted to use one to call a cousin. The code was old, and the operator interrupted the call and asked my cousin to pay for it. That ended my career as a phreak. Around the same time, give or take a year, I had two extraordinary phone experiences. One night I was chatting with a friend, Barbara, and out of no-where we were connected to a men's prison. One voice after another got on the line, the men as amazed to be talking to pre-teen girls as we were to be talking to convicts. Another time I picked up my phone and heard my best friend and my boyfriend talking and heard him ask her to the prom. I listened for a while, until she said yes, and then exploded with rage.

Years later I realized I must have known a phone phreak who never confessed his skills to me. Perhaps he will see this and grin. I did spend a great deal of time in high school calling a phone number that connected me to a phone line that dozens of other people were on at the same time. I as-sume now that these were phone phreaks, but I didn't have any clue then. I don't remember who I spoke to, or about what, or even who gave me the number, but I do know these calls were some of the most interesting ones I ever partici-pated in. These days the general public finds this type of con-versation in 976 phone numbers, but they are for-profit lines and contain people conversing solely to get sex and/or dates.

More than a decade later I got an Atari computer but missed the golden age of modeming because I was too busy playing *Centipede* and *Space Invaders*. I was also busy with college, graduate school and work. In 1991 I got an IBM-compatible computer to begin publishing a Yippie-type maga-zine of my own called *Gray Areas*. The magazine doesn't publish information on how to commit crimes, but it does legitimize the people who commit them and explain how and why people break the law. I did a stint calling BBSs because I wanted to explore how software spread and because I loved the idea of chatting on them. I met a number of local hackers (without knowing it) and must confess that when one tried to talk to me after telling me he was a hacker, I ran in fear. It's not possible to be around the computer underground very long without having some regrets, and the fact that I did not get to know this hacker then is surely one of mine.

The first issue of *Gray Areas* was reviewed in *Computer Underground Digest* and *Phrack,* and through those reviews people in the computer underground contacted me. One of them was a virus writer, and another was a former *Phrack* publisher. I talked to them both "voice" for hours at a time and learned enough to know I wanted to learn a lot more.

In July 1993 I found Internet Relay Chat (IRC) and thought it a stupendous forum for communication. One night one of them called me and I mentioned I had been there. He told me about the IRC channels #hack and #phreak, and I checked them both out the next day. Channels in IRC are topic-based chat forums where specific people often hang out. At the time, these channels were secret and one needed an introduction in order to find them. These days #hack is mentioned in dozens of books on how to find cool things on the Internet and thus I feel comfortable writing about it because its presence on the net is no longer the big secret it once was.

Hackers on IRC

Internet Relay Chat, a real-time chat facility divided into "topics" by channel, is one of the main hangouts for hackers. Although IRC is huge (up to nine thousand people from all over the world chatting at the same time) and only a few channels out of hundreds talk about hacking or even computers (channels include #poetry, #hottub, #disney, #texas, #gaysex, and so on), it must be noted that, for the most part, the hackers are the bad guys throughout IRC, taking over the other channels and making IRC unpleasant for almost all who enter. It's fair to say that the permanent removal of fewer than one hundred hackers would make IRC and the net in general much, much safer and more pleasant. But hackers defend their actions by arguing that the net would be boring without them.

IRC is accessible across the Internet on machines with the right software. Anonymous handles conceal real identities, so no one has any idea whom they are really talking to on IRC; because of this people come into contact with hackers and end up hurt. It is possible to get on IRC, encounter a hacker on the prowl for someone to abuse, say the wrong thing and end up with your account deleted in the space of twenty minutes. Or less.

To illustrate how complex this gets, hackers can log in to the #hack channel using software (like *screen* or *UW*) that allows them to come in from several sites and be on IRC as many separate connections, appearing to be different people. One of these identities might then message you privately as a friend while another is being cruel to you in public. Or they might set up fights with their other identities on the channel to see who sides with whom. Since people tell different things to different people, it makes sense that multiple identities can pull more information from others than just one.

Another exercise of hacker power is to hack channel operator status. Operators have the power to ban abusive IRC users. The IRC operators, however, often screw with users themselves, thereby causing the very problem the position was intended to prevent. Hackers want to be "ops" on IRC for reasons ranging from keeping people they don't like out of a channel they are on, to taking over the channel and removing the people on it, just to harass people. Some people's egos are so attached to having ops privileges at all times that they get on IRC and literally beg to be blessed with ops in order to have power over those without it.

IRC and the computer underground in general are fluid anarchies where power shifts from day to day and hour to hour. You become instantly "elite" depending on who your friends are, what item you possess that everyone wants a copy of, or whom you just humiliated. Most of the hackers who hang out on IRC are unskilled ("lamers") and can only use a red box someone else has built for them, or hack root only by using someone else's tools and information about known bugs. ("Root" means having the password to the root account, which has permission to do anything in UNIX.) They get considered "hackers" either by hanging around long enough to achieve seniority, by breaking the law in front of key people or by acting tough and picking on people so that no one will notice they have no skills.

Once, having been given a cool T-shirt that said "Net.God," I expressed some reservations about wearing it in front of other hackers. I asked the few I knew whose stolen trophies proved they most deserved the title, and they each said something to the effect that I should feel free to wear the shirt because almost everyone else wearing it knew not much more than I did. One of them also told me that there were all kinds of hackers and that I was a hacker because I hack with words.

Joining #hack

I went to IRC originally from an account on the WELL that had my handle as "grayarea" and my real name visible. I had no way of knowing it then, but at the time more than a dozen regulars on #hack had root on the WELL or had been given access to one of the several back doors (illegal entrances to the system installed by hackers). These regulars thought I might be a system administrator or a Fed investigating a break-in and read my email to see who I was; one of the people who read my mail was so taken by it he chose to share the mail with his colleagues on the channel. I had no idea email could be so easily stolen or that so many people had access to mine. When the WELL called me to say I had been hacked, I made it my business to find the people who did it and to learn why. In the long run it made my career, both because of the article I ended up writing about it in *Gray Areas*, which greatly increased our subscriptions, and because many of the hackers who originally spied on me became so curious that they eventually wanted to talk to me as themselves, using their real names and/or real "handles" (nicknames used to hide one's real identity) as opposed to fake handles created just to fool me.

Hackers are notoriously secretive people, and thus the friendships I ultimately made from the WELL break-in could not have been bought for any amount of money. But there were weeks of being "banned" (locked out) from #hack and being "killed" (having your login go dead so you have to join IRC all over again) and being asked questions about my intentions regarding the WELL story. I don't want to minimize how ugly some of these people were in their attempts to scare me away from writing anything or in trying to make me fear them if I did write anything.

I have been to hell and back with the two hackers who consented to interviews about the break-in. Their trust was the easiest to win initially, and that trust has been tested repeatedly as law enforcement has since questioned me about their identities; we have been in and out of contact as they have felt secure or paranoid at various times over the years. Our values and expectations of each other didn't always mesh, but despite that, it's one of the highlights of my life that I was able to find two hackers the WELL and law enforcement could not. The positive things I say about hackers are the result of their being willing to take the time to explain to me why I was hacked and how to arm myself better against hackers. I regret not interviewing the third hacker who offered to be interviewed for that article. I told him, "Nah, I got enough." Boy, was I ever wrong. Hackers have different motives for entering sites, and they also each interact differently with me. Since I often bond with my interviewees, I blew a chance to get to know someone I could have learned a great deal from.

Once I joined the hacking channels, it took about seventy-two hours for word to spread that I was a female reporter from a magazine none of the hackers had ever heard of (90 percent knew within eight hours or less of my initial arrival). Once it was clear I was not a WELL administrator after them for the break-in, that I intended to stay, and that my email revealed I talked to hackers who hate each other, knew hacker secrets, and thought hackers had personality problems, people started to try to get to know me. You'd think that would simply mean conversation, and of course it did, but unless you have some personal experience on IRC or with being socially engineered by hackers, you can't fully understand. Some of the hackers stole my trash, pulled my credit reports and, although no one used them, teased me about the contents just to prove they had them; they hacked my other net accounts so all I did and said was watched; they called me pretending to be salesmen to see if I would give them credit card numbers, or to see whom I lived with, or what hours I kept, or just to hear my voice. They forwarded my phone number and then disconnected it as well as repeatedly putting a "maintenance busy" signal on my parents' phone line. They set up "bots" (remotely controlled IRC scripts that make it appear you are talking to live users) who befriended me on IRC and coerced me into pouring out my innermost feelings while the real users laughed out loud; they asked me question after question and when they got the answers they wanted (or not) were never to be heard from again. They deleted my email, preventing articles, advertisers and important business messages from reaching me. They conference-called my answering machine with insults, threats and outright lies. Even when they lied to me, I responded with the truth, and sometimes neither of us knew what to do then. I was probably the most honest, open person ever to come to #hack and I got my teeth kicked in for it on an hourly basis. I was an embarrassment of riches to a malicious teenager. I cried and admitted it to them. I even cried in public when I spoke at PumpCon 1993 and told the audience that the harassment of my company had gotten so bad I might have to turn to the FBI for help.

As I know now, this stuff happens to almost every newcomer in the underground (but rarely do they have a corporation at stake), and some people come to IRC not to converse but to ruin the day of other people they happen to stumble into there.

Things changed after I spoke at PumpCon because instead of the good hackers compensating for the bad, most of them ran away that day (or that night when they heard about it on IRC); some never spoke to me again. Alas, more hackers started picking on me because it is taboo to show weakness—I had committed the ultimate sin. I had also threatened to "narc," which no one could forgive. Little did I know at the time how uninterested law enforcement would have been in my situation. There isn't anyone in this community who isn't hated by someone else in it, and if I had just not let it show how much I was being hurt, lots of what they did to me might never have happened.

The Hacker Women

Not that many women frequent the underground, and most that do come into it as transients while they are dating a hacker or as press to do a one-time story. On IRC, about two dozen women hang out with hackers regularly; I'd imagine three times that many go to *2600* meetings or to Cons but do not frequent IRC. I have met more than a thousand male hackers in person but less than a dozen of the women.

Women are accepted to a point in the community. Some hackers like to talk to women even if it is not about technical subjects. They befriend all the "girlies" and try to have an important place in each one's life. Other hackers bond with one or two of the women and think the others are psycho. Even those hackers that claim to hate women end up talking to them (or at least being civil) if one dates a pal, so no generalization is absolute. At times the men get in a mood and kick all the women off the channel, supposedly so they can talk about the technicalities of hacking, which they feel the women aren't truly interested in; but this never lasts long, because the men can't talk too openly about their hacking exploits with all the narcs and feds around.

Generally the women are there as groupies. There's even a T-shirt that says "Fuck me, I'm in LOD," referring to the elite hacking group Legion of Doom. Even the most professional of the women has supposedly slept with a hacker or two. Considering that this is where the women spend the bulk of their free time, and how many men there are per female, this is not all that surprising.

A few of the women seem to have no enemies or at least none that stalk them. However, that is the exception rather than the rule. Most of the women have been stalked on- and off-line by at least one hacker, and some of them are regularly harassed in real life by several. One woman had her phone turned off by a hacker she would not have sex with. Another woman had sex with a hacker, and they hit it off so badly that the police and her university got involved more than once.

I've had a lot of problems with harassment, as I said, partly because I'm a journalist. I attracted the wrath of an entire group of phreaks whose members are based across the United States as well as in Canada and Scotland. They used the phone number for Gray Areas, Inc. as the example of how to charge your calls to someone else in an electronic zine available across the net on FTP and World Wide Web sites. I called the main site at a university the leader was operating from and got absolutely no help or understanding. After seeing that no one cared what they did, they hacked my main net account and used the information they pulled from there to launch a fake electronic version of *Gray Areas.*

Another hacker who works for a large net provider took it upon himself to ban me from the #hack channel and to call me a child molester, prostitute, drug dealer and other outright lies. He tried so hard to ruin my reputation (even picking a fight in real life at HoHoCon, which backfired on him) that finally I felt I had no choice but to write about it and make an example of him. I called his company to verify the hours he worked and his title. Although the personnel director took me very seriously, when word got to his boss she took him at his word that he never IRC'd from work.

Not only do sites not give a damn what happens to net users unless they get their files removed, many of the top Internet providers, like that one, also (knowingly or unknowingly) hire hackers to interact with their customers. The WELL would not hire one hacker as their system administrator because he had served jail time; instead they hired a former member of a hacking group, and the WELL ended up hacked by people the sysadmin knew quite well from when he was actively hacking. It's a catch-22 situation though, because for the most part, only a hacker or former hacker cares enough to tinker with fixing all of the latest bugs in the operating system. UNIX, for example, has *lots* of security holes that hackers can exploit if they aren't patched.

Drawing Ethical Lines

No one who hangs out with hackers has been killed by them yet, but it should be noted that most hackers own guns or aspire to (even the minors). Because of their age most hackers are in decent physical shape to begin with, yet a surprising number of them also study martial arts, work out and lift weights. I believe that hackers often hack because they desire power and control and that they build up their bodies both to be able to exert physical power over others and to be able to defend themselves in prison, which some of them accept will likely be a part of their lives. Physical violence (both online and in real life) is threatened a great deal for the stupidest of reasons, like an insult uttered on #hack to someone who was in a bad mood. Less physical responses are common too; hackers may take down entire sites in anger at one user or may just delete the contents of that person's account with a note to the sysadmin to delete the user or they will return and do worse. These incidents are shocking and malicious, but responses like these seem to many hackers to be the only way to make a point after they've been really pissed off.

You might think that hackers would not prey on their own, but some of them prefer attacking their peer group rather than targeting strangers. Alas, there is no hacker's code of ethics other than, perhaps, "We hack because we can" or "Hack him before he can hack you." The women, especially, are seen as easy targets because they often lack the skills to fight back and will generally take more before they ask for help. Being nice and apologizing if you make an error simply does not cut it in the computer underground. Your only hope is if someone more skilled than you comes along, takes sides, and chooses to assist you instead of them. And hackers don't seem to stop, no matter what you do to try to make peace. Violence just begets more violence, but it is the only response some of them understand. As I write this, the hacker who called me a child molester stopped bothering me for a few months after receiving an ugly voice call at work (which I didn't know about until after the fact) from some hackers who called *him* a child molester; they made it clear they could provide proof to his boss that he indeed IRC'd and harassed me from work.

Sites get hacked for lots of reasons. A hacker may simply want to read the email of a female who uses that site, or if he's a minor, to get local access to the net that he can't afford or purchase without parental consent. Hackers may break into a site to learn the operating system, to get the phone number of a hacker who has a legitimate account there, or to use that site as a base to hack out of or (as Kevin Mitnick did on the WELL) as a place to hide files they stole from elsewhere. One hacker broke into a university site (which he revisits regularly) so he could raid the account of a user who is a courier for a warez[4] group and get all the newest pirated software the instant the courier gets it to distribute.

There are different types of hackers, too, drawing very wavy lines about what they will and won't do. Some of them claim to draw the line at accepting money for their services. Most of them change their tune when someone actually offers them real cash. Some of them draw the line at deleting files, claiming that as long as they just look at files or copy them the site is not truly injured. Over time though, they often end up removing the files of a user, admin or entire site either because they lose their temper or because they find something on a site they feel should not be there. There are hackers who dread breaking the law and hackers who are so criminally oriented they dread complying with it. There are hackers who would never go after an individual, and others who would rather know your Social Security number than know you.

Some of the wide variety of exploits hackers claim to

have pulled off include: releasing a prisoner from jail; compromising an anonymous remailing service in Finland; playing with the phone lines at a radio station to win automobiles; installing fake ATM machines in shopping malls; stealing and distributing the credit card information of more than twenty thousand users of the net provider Netcom; interrupting phone service and getting the unlisted numbers of celebrities from credit bureau reports or by sweet-talking a person at the phone company out of them, or by determining an artist's manager and pulling phone records on their line. It becomes clear, finally, that ethics don't exist in this community because hackers never know what they will stumble across next and once something is perceived as *possible* there will always be those with the urge to accomplish it.

The Nature of Friendships

So why do I stay and even defend hackers? First of all, they are not all bad. Many of them have hearts of gold and wouldn't hurt a fly. Some of them play with computer and phone systems but never with individuals. Some of them are mean to other people but are really nice to me. Some of them are good at kissing up or making it seem as if they like me to my face, and even though they might badmouth me as soon as I log off or walk away, in all the time I have known them they were good enough at it that I haven't seen through them yet. When they turn the charm on, or talk to you for five hours and challenge your brain, or tell you stories about their exploits, they can be the most interesting people you'll ever talk to. In fact, after a while non-hackers pale by comparison.

Once I met with a member of a virus-writing group and got so involved in the conversation that I forgot I owned a car whose meter needed to be fed. I got a fifty-five dollar parking ticket while sitting in McDonald's listening to him talk. Thirsty at one point, I went to buy a drink but when I came back with it, I noticed I already had two other drinks sitting untouched on the tray! Another time I had a blind date with a hacker that went so well I ended up bringing him home to stay with me for five weeks. The conversation was simply that good.

I don't want to imply that the friendships are lifelong because at any moment these people can and do get busted and are pulled away from you. Or you try to get too close and ask the wrong one to spend the night with you, or mumble that you love them as a friend, and it ruins the "friendship" forever. Or you lean on them once too often to be unbanned from an IRC channel and they get sick of you. Or you find out they have been lying to you the whole time you have known them, that you rejected them once under a different handle, and now they've resurfaced with a new one so they can keep hanging around you. Part of the problem I have relating to them is their young age and lack of life experience, part of it is our different reasons for being in the underground, and part of it is surely that I keep expecting them to give of themselves as well as receive. You may get flowers or a card from a hacker on Valentine's Day or for Christmas or your birthday, but it's not likely to be from the ones you care about the most. And then there are those who befriend you simply to get something from you, and once they get it they move on.

The Future for Hacking

What can society expect in the future? Certainly no tolerance from this group that draws its strength from being misfits. Ninety percent of the hackers are not "lifers," so we can expect new ones to arrive and most of the old ones to go when they get a job they care about or a girlfriend who sucks up their time. As with most deviant groups, the lifers are usually the ones who have arrest records or are under active investigation. They are the most committed to the criminal lifestyle, and they have often already been cast as deviants by the legal system or by the media and therefore feel they might as well continue the behavior. Law enforcement agents will always be far behind in catching hackers as the agencies lack the technical knowledge to understand how the crimes occur and lack the budget to arm themselves. Attempts at hiring hackers within law enforcement have for the most part failed; and it's a tough assignment to catch people who thrive on being invisible and untraceable.

The reality is that a hacker can do anything he wants to anyone and will most likely not be prosecuted for it. For the most part, no one wants to hear about it. In fact, many sysadmins hate users who complain more than they hate hackers. Even if someone does come after a hacker, in general all that happens is his computer is seized; no charges are usually filed. Jail sentences of six months to one year are given out rarely, and as of this writing the most time a hacker has ever spent behind bars is four years. Penalties will probably increase eventually, but at present you are in far more trouble if you are caught with a few hits of LSD in your pocket at a concert than if you stalk someone, turn off their phone and impersonate them electronically.

As a result, hackers walk around believing they can do anything they want as long as they are willing to risk losing their computer for it. Community service, one frequent sentence, is a joke. Most of the hackers I have spoken with had someone sign off for their hours after they had worked far less than the amount required, and the rest generally get work with computers anyway. Convicted hackers could be put on probation and sentenced to secure the sites they broke into; but, of course, this leaves too much power over the system at the discretion of the hacker, and to many, this would be seen as fun, not a punishment.

Instead of continuing to use the same methods for punishment (which clearly don't work), someone should study hackers as I have to see what makes them hack, and consider that when deciding on criminal sentences. What most of them want is some legitimate job working with computers. Unfortunately, most of them have such low self-esteem that they have decided they will not get one and start hacking before they are even of age to try to work legitimately. They seem to have a couple of characteristics in common as a group: many have absent fathers or, at least, parents who don't notice what they are doing; a few have been placed in mental institutions or sent in for counseling; some suffer from severe depression or hyperactivity and may take medication for it; and most have explosive tempers and/or egos that need to be continuously fed. They are the classroom bullies who come home to turn on their computers and become even more powerful by virtue of their anonymity. Or if they aren't bullies by day, they become bullies when they get on the net.

Gray Areas covers everything from drugs to prostitution to bootleg tapes to UFOs, yet the computer underground is the grayest area the magazine explores. It is also the one that takes the most out of me personally, draining me so badly some days (or weeks) that I cannot deal with anything else.

But the increasing use of computers and the Internet surely warrants a publication that deals with this netherworld.

From experience, my best advice is to use *PGP*, or *Pretty Good Privacy*, a publicly available file encryption program, on all email and to encrypt and decrypt messages off-line. You are not likely to attract the interest of hackers unless you put yourself in their faces somehow as I did. I hope for your sake that if you do choose to interact with hackers that you come to care less about them than I have. To quote Carole King, "They'll take your soul if you let them."

Endnotes
. . . .

1. Computer wizards who manipulate computer systems for reasons ranging from innocent exploration to malicious destruction to fraud.

2. FSP is like FTP, but uses UDP connections instead, so files can be sent and received through firewalls.

3. A script is a program which automates computer commands, allowing users to save time by just running the script and sometimes allowing them to run commands they do not fully understand.

4. Warez groups are groups of hackers who steal programs (most often new computer games) from the computers of game companies and make them available on bulletin boards.

We Are Geeks, and We Are Not Guys: The Systers Mailing List

L. Jean Camp

> A ship in port is safe, but that is not what ships are for. Sail out to sea and do new things.
> — Admiral Grace Hopper, computer pioneer

The Internet can be rough sailing for women, buffeted by the high winds of derision, sucked down into whirlpools of contention. But in that sea of bytes there are destinations beyond compare, worlds to explore. And as we steer our craft out to sea, braving the great unknown, we know that in this world there are electronic ports where we can become refreshed, refueled and ready to sail again when we may think ourselves alone.

Systers is one of those ports of call. Systers is a mailing list of women in computer science and related disciplines. We are geeks, and we are not guys. Not guys, but geeks! How can that be? But we are, we have been and we continue to be. If it surprises you to learn that more than fifteen hundred feminist geeks are out there, imagine the surprise to each of us!

Being a geek isn't easy. It's hard, intellectually challenging work. For some people, people whose gender I won't go into here, people of less intellectual capability than some of my systers, achieving true geekdom means sacrificing their emotional development. You might have met some of these people. Technical universities and work environments tend to be full of them, and are brusque, competitive places as a result. My place of work, Carnegie Mellon University, is no exception.

So is it any wonder that I turn to the net in search of solace? But there I find that all the groups formed for women quickly become swamps of men's bile. One man told me, "I know as much about being a woman as you do." After all, he lives with women, he probably has been intimate with some women and he spends so much time thinking about our many flaws! He certainly knows all about women.

Even the discussion groups that focus primarily on parenting have become arenas for men to pat themselves on their collective backs, to discuss how much more difficult it is to be a father than a mother, and to discuss the discrimination against and oppression of fathers. There is no end to the complaints of the anxious and oppressed white male on the Internet.

Consider a Usenet newsgroup specifically started to discuss issues about women: *soc.women*, where the posts by men outnumber the women's. On the Internet, as in life, men dominate discussions about women. Many of the feminists on these newgroups are indomitable warriors; there will always be a battle for them. But some of us chose to spend our energies elsewhere, and even indomitable warriors need a place to rest.

Too often, when women try to create spaces to define ourselves, we are drowned out by the voices of men who cannot sit quietly and listen, but need to bring themselves into the discussion. Many of these men support women. But the voices of men who cannot be silent even in a space ostensibly devoted to women means that there are no public spaces for women to talk about and to other women.

So we withdraw to a room of our own—to mailing lists. Even the most indomitable woman needs a port of call. Here we chatter and type in nurturing communion, knowing that the world cannot do without our unique contributions.

Systers' History

A mailing list is a list of email addresses kept on one computer. A message directed to the list goes to that computer and then is automatically copied to everyone on the list. Someone subscribed to a mailing list generally finds her mailbox full of an endless stream of messages full of earnest discussion, gossip, jokes and occasional discord. Each message by a mailing list member spawns its own replies, a round-robin discussion that at times resembles a support group and at other times a graduate seminar. A bad mailing list can be dull. A good one can be wonderful.

Systers, my port in times of storm, my destination of choice for R&R, was begun by Anita Borg (see her account of Systers on page 142). Dr. Borg is a senior computer scientist at Digital Equipment Corporation (being senior at DEC is a big fat hairy deal in the computing world), a pilot and the benevolent matriarch of fifteen hundred systers. She is, as she has named herself, her systers' keeper. Systers is an unmoderated but strongly guided mailing list open to women only. To join, you have to swear you are a woman. If it turns out otherwise, you are removed.

Moderated lists have moderators who sift and sort the stream of messages, culling here, compiling there, much as a good hostess directs the flow of traffic at a dinner party. The default is that all messages go out to everyone. If a discussion begins to dominate, posts made in reply to that discussion, called a thread, are not sent out to the list. Moderation in all things. When Dr. Borg guides mailing lists, she often posts only *Cut it out*. This usually works. She does not view every message before it goes out, as a true moderator does.

On Systers, a woman might send a question or comment to the list and then a volunteer, often the original poster, offers to summarize the results. This prevents list members' personal mailboxes from being flooded. Sometimes a post will cause a flurry of responses: Take Our Daughters to Work Day, for example, was much discussed. Pornography, needless to say, is an issue not to be mentioned under the threat of a firestorm of passionate debate and many resulting "unsubscribe" messages.

Note the discussions in *soc.men* and *soc.women* that relate to women: Threads there include discussions of why men are smarter than women, as proven by SAT math scores.

Men arguing against maternity leave since women choose to be pregnant. Men who want men to be able to choose whether or not to admit paternity ("choice for men"). Why men make better parents. Abortion. Abortion. Abortion. How men take all the risks in dating since they have to ask for the dates.

The discussions about women on Usenet are just that—about women. Not by women. About.

Consider the topics that women talk about on Systers when *we* control the debate: how to recruit more women to science. How to deal with illegal questions on interviews. When/if to have children. Whether or not to go to grad school. How to deal with a coworker who harasses you. How to deal with email harassment. How to deal with a job hunt when there are two careers involved. What to do about childcare at conferences. How to select an advisor. What effect would the selection of C rather than LISP or Scheme for a first programming language have upon women in computer science? What good fellowships exist for women?

For even more specific discussions, Systers has sublists for specific affinity groups: for example, for women of color and for lesbian and bisexual women. How freeing to have a place where they don't have to deal with people very unlike them discussing "reverse discrimination" and "special rights" when they want to talk about how to just get through the week.

Systers Didn't Let Me Down

It was early in the morning and I knew it would be a very long day. I had a paper to present at a conference and not one but two ear infections. The conference—The Telecommunications Policy Research Conference—is the major one in my area of research, and I had twelve hours to finish my paper, get in the car and drive there to present it. My advisor had stayed up until the wee hours the previous night helping me polish. I was still editing.

My husband was actually going to do the driving, so I could work on the presentation in the car. He was coming with me because of the baby. Five months old. Still breast-feeding. Squiggly and helpless and wonderful and everything unprofessional in the world.

Going to a conference breast-feeding a baby? There's no

chapter on "Dealing with Let-down in Silk" in *Dress for Success*. I didn't know what to expect, and I didn't have anyone to clue me in. So that morning I sent out a message, a quiet cry for support to one thousand other professional women in computer science.

They heard me. They took my virtual hand, they gave me virtual hugs. I was not standing in the wilderness. Trailblazer? Hell no! So many women had gone before me, lactating their way through dissertation defenses, conference presentations and teaching tutorials to top management that it was no impenetrable forest I stood in, but a clear road with a clear sign marked not "This Way," but "It's Okay."

My message that morning was contained. You could hardly tell I was holding my breath:

I am going to present at a conference this weekend. Since I am breast-feeding I am taking little Addie along (4 mos.) My husband is coming along, too, to take care of Addie during the day (wonderful?—yep! he is). If you have taken a baby to a conference I would really like to hear your experiences. I feel like a stranger going to a mostly male conference w/ baby. How did you handle it? What problem did you have—what did you avoid with good judgment calls? What was most helpful/worst?

The responses flowed in. One. Two. Six. Dozens. Without Systers I would have been astoundingly alone—how many lactating technical doctoral candidates do you know? Instead, I was comforted, told that I, a normal soul in a body leaking milk, could handle it. I felt like a pioneer, going into the great unknown when I sent my note to Systers. When I left for the conference, I had had dozens of responses and I knew that I was not (and never had been) alone. I got messages from women who had gone to conferences the previous month and women who had gone fifteen years ago! One woman told of taking her five-week-old baby to a conference and she was preparing, seven weeks later, to take him along again. They all said, "You can do it. You will be fine."

The one message I did *not* get was: It was terrible and destroyed my career. The horror stories helped there. One woman's little one did one of those massive explosive poops that violate all the laws of physics right in the middle of a session. She readjusted the baby carrier and went to the motel room and cleaned them both up. If their babies could be sick or have an ebm (e = explosive) in a session and all turned out fine, then, well, what was I so worried about!

When I returned from the conference, I reported back to all the women who had reached out to me:

Knowing that there is one incredible person is not always so helpful, simply because this person is so exceptional. But knowing that there are dozens of women, that I did not have to be so unique or incredible, made me feel like I, too, could pull though.

And after all those messages the one thing I was not when I left was so worried. Thank you all. Things went well for me at the conference. I ended up encumbered not by the baby—but by my ear infection!

The Lighter Side of Systers

Systers also serves up lighter fare. Feminists often get abuse for not laughing at funny jokes. Maybe it's that the jokes we're told aren't all that funny. One syster was sent a bunch of jokes about engineers, all of which assumed that engineers are male, for example: "You know you're an engineer when you have a beard because you have calculated your efficiency loss in time shaving and found it unacceptable."

This syster considered sending out a flame: "Hello! Remember me? The engineer? The woman?" Instead she got onto Systers and asked for jokes that assume engineers are women. The result:

You know you're an engineer when...
—you have hairy legs not as a political statement but because you have calculated your efficiency loss in time shaving and found it unacceptable.
—they give you drugs during labor not because you can't take the pain but because you keep trying to rebuild your monitors.
—you try to register in the automotive department for your wedding gifts.
—you are excited about your first period because it gives you the opportunity to test the viscosity meter in your chemistry set on an interesting biological sample.

Instead of flaming the men for their sexism, she sent back a collection of jokes from our point of view. Who says we don't have a sense of humor?

The Barbie Experience

Systers is a powerful personal resource for women, but there is an important public element, too. Nothing illustrates this better than the great Barbie fracas. If you recall, Mattel introduced a Talking Barbie in 1993. This Barbie said things guaranteed to appeal to the mostly brain-dead. Among the gems that sprang from her perfect lips was "Math is hard."

An alert syster, possibly the appalled mother of the owner of one of these dolls, sent out a message. After all the discussions we'd had of how to keep young women and girls interested in math, this was a broadside.

Women study math because mathematical competence leads to more money, which leads to many good things, such as autonomy. Math is fun. Math is good. Math and technical knowledge are power. Here was Mattel saying clearly to girls: Stay away from math. You are not interested in any quantitative professional career. Take domestic science! Be dependent.

I read the *Washington Post* and skim the *New York Times* and the *Wall Street Journal* regularly. I have The Associated Press and United Press International wires available to me via Clarinet. But Systers spoke first.

Actually, first we *screamed*. Then we discussed it. Then we got down to business. Never mess with systers.

We found the number for Mattel's complaint line and started calling. Individual systers called their professional organizations to complain, and the American Association of University Women did just that. The mainstream media picked it up—after Systers had begun the battle.

Mattel surrendered: Barbie no longer advocates female innumeracy. Systers got women together, and we acted.

Join the Fray

Computers can give you a level of anonymity, which may give some women who've never spoken up in public the courage to express themselves. But anonymity is isolation, a level of invisibility. After playing around Web pages and lurking on Usenet groups, you want to be seen and to define yourself. You do not want people to define you, especially not by simply looking at your name and guessing your gender.

Systers has given me comfort when I needed it, reminding me every day that I am not alone. The feeling is small, but constant. As Systers has filtered into my being over time, it has become a tremendous positive force in my life. Not being alone means not being hidden. Of course, Systers speaks to me as a technical woman in academia. But there are many other mailing lists for women—no woman needs to be alone on the net.

The very strength that Systers offers can make it a sanctuary on a hostile net. But we cannot live in a sanctuary, regardless of the temptation. It is important to go back out into the public debate and remain visible, if for no other reason than to ensure that no woman is left truly isolated. The power of connectivity to effect change is truly incredible. The Barbie experience taught me that. The far right realizes it, too, and is organizing on the net. We need to be doing the same. If not, it will be as if women were sending out missives via caravan while those who would deny women their rights were using Cruise missiles. If we're not there, the doors of electronic democracy will be closed to us.

Dr. Anita Borg on the Mailing List Systers, from *Computing Research News,* September 1994

The existence of exclusively female forums is controversial and legitimately so. Exclusive forums such as male-only or white-only or Christian-only clubs have been used to exclude other groups from information and power sharing. As the founder of Systers, a large female-only mailing list, I have frequently been called upon to justify the exclusion of men and to explain why Systers is not discriminatory in the above sense. This article attempts such an explanation. I hope to generate discussion, but more importantly, to generate understanding and cooperation.

Increasing the number of women in computer science and making the environments in which women work more conducive to their continued participation in the field require the active development of both women and men. In particular, there must be ongoing and productive communication throughout the field concerning the unique problems that women face when they enter the field and as they progress and advance. The fact that women are a small minority in the field results in two impediments to this communication. First, women work almost exclusively with men and so have few opportunities to create and then participate in a "community of women in computer science." Second, men work almost exclusively with men and have limited opportunities to communicate with more than a few professional women. Open electronic forums can improve communication by introducing us to a larger community, but do nothing to reduce the disparity in numbers. On the other hand, exclusively female forums, such as Systers, are a particularly effective way to connect women in our field with each other. They also ultimately contribute to improved communication between women and men.

Let me first describe what Systers is and what it is not. Systers is a private, unmoderated, but strongly guided, mailing list with a documented set of rules for participation. The mailing list includes female computer professionals in the commercial, academic and government worlds as well as female graduate and undergraduate computer science and computer engineering students. Systers currently has over 1,500 members in seventeen countries. We are a global community of individuals who are otherwise physically isolated from each other.

Systers is a civilized and cooperative forum in which "flaming" is rare and personal attacks are actively discouraged. We ask that Systers mail not be forwarded nor its contents used outside the list without the permission of the contributors to a message. There is no rule of secrecy in Systers. This rule simply empowers our members and protects our privacy by giving each of us control over the breadth of distribution of our comments. It is based on a common courtesy that, if applied more generally, would make the net a more hospitable place for substantive group problem solving.

Systers is not analogous to a private all-male club. It is different because women in computer science are a small minority of the community. It is different because Systers is not interested in secrecy or in keeping useful information from the rest of the community. In fact, useful messages are regularly made public after checking with the contributors. The likelihood that an underempowered minority will keep inaccessible information from the large empowered majority with every means of communication available to it is small indeed. I have not addressed whether a forum such as Systers would be necessary in an ideal and egalitarian world or even in a world similar to our own but with many more women in computing. When we get there, we can make that decision.

The following paragraphs enumerate the reasons for keeping Systers a female-only forum. None of these benefits accrue to women in other existing open forums. Women need a place to find each other. As a geographically dispersed and frequently individually isolated minority within computer science, women rarely have the opportunity to interact in person with other women in computer science on any subject. Women (and men) have many opportunities to interact with men. Until Systers came into existence, the notion of a global "community of women in computer science" did not exist.

Women need female role models and mentors. A primary function of women-only interaction is mentoring. Exposing women to the full range of significant interactions among women, without the perception of help or input from men, serves to bolster self-esteem and independence. This includes exposure to women discussing purely technical issues among themselves and shows that this makes women more rather than less able to interact professionally with men.

Women need a place to discuss our issues. Many open forums whose focus is women's issues suffer from a common problem. Discussions are frequently dominated by disagreements between men and women about what the issues are rather than how to deal with them. This is not a problem with all men, but is a problem with almost all such open forums. Women more often share common ground that allows us to get beyond defining issues and on to constructing solutions.

Women need to discover our own voice. Discussion among women is different from that of women together with men. Men, even when in a minority and even when well-meaning, have a different style of interaction. They often dominate discussions. Even when they don't, the style of a mixed conversation tends to be in the style of male-dominated discussions. As women understand more clearly what those differences are and what professional discourse is like on our own, we will be better able to bring our voice to open forums.

I recently received two messages that illustrate how Systers helps women participate more effectively and more professionally with men.

A researcher from an industrial lab stated, "When I first joined the list a few years ago, I was skeptical about the need for a list specifically devoted to issues facing women working in computer science. But since then, I have become much more aware of the differences in the ways men and women interact, and many of the experiences and views shared by others on this list have helped me to better understand how to function effectively in a male-dominated research environment." A university professor described a change in her students: "The availability of the list to our women graduate students here at [the university] has had a remarkable effect on our students. The women are becoming more self-confident and more aggressive in their dealings with our male-dominated faculty, many of whom still regard women as out of place in the program."

Systers is definitely not the only forum in which concerned women participate. It is only a starting place and place of respite in our journey to equality. It is essential that we continue to actively communicate and participate with men, that we not become isolated from professional men, and that we bring our issues to the fore at every appropriate opportunity. Since most of us work exclusively or nearly exclusively with men, it is actually impossible for us to become isolated from men even if we wish to be. Since men make up the vast majority of the field, it would be foolish to believe that real change could take place without them.

To include men in Systers would take away a vital source of mutual support from women. On the other hand, the need for serious discussion in an open forum exists. It behooves whoever runs such a forum to realize that women who have experienced conversation on Systers will be for the most part uninterested in participating in a wide-open free-for-all. The commonly applied list-management principle "if you can't take the heat, get off the list" will not work. It has been tried and has failed. The forum will need a strong leader/moderator, committed to the encouragement of productive discussion and willing to stop unproductive argument. I do this for Systers. While I do not have the desire nor the energy to run another forum, I am surely not the only person capable of it and offer my help and experience to anyone who is willing to take on the task.

It is not the reluctance of women nor our participation in forums like Systers that limits communication and joint problem solving with men. It is the sexism in our society, our field and our consciousness that limits us all. If men want an open forum and are seriously interested in hearing what women have to say, rather than in telling us what we need, then such a forum could be a fruitful and productive sibling for Systers.

Not for the Faint of Heart: Contemplations on Usenet

Judy Anderson "yduJ"

While attending Stanford University, I witnessed the birth of the Internet, and have since watched its growth spurts and growing pains as connectivity has increased. I got my first email account in 1979 and joined my first netwide mailing list in 1980—in other words, I have seen a lot of different styles of email, mailing list and netnews messages. It has not been my perception that my gender was relevant to my participation in this new medium of computer-mediated communication: I offer you my thoughts, experiences and advice on the net as a *person*, not just as a "woman online."

Me and Ten Guys

I grew up in the heyday of Women's Lib. I didn't pay a whole lot of attention to it, really, I just went along with my life one day at a time. But my mother never said, "You can't do that—

you're a girl," so I played with model railroads alongside my Barbie dolls, and felt no aversion to math and science classes. When I was in high school, I started to notice that most girls weren't like me. I still had a few female friends, though, so it wasn't blindingly obvious; enough men were also sufficiently different from me that I did not draw deep conclusions along gender lines. But by the time I got to college, girls (now women) had become an alien species. They cared about things I had no interest in, and vice versa. Hardly any of them were science fiction fans, and those that were preferred fantasy. I hung out at the computer center, and most of the time when a group of us would head out to grab a snack, it was me and ten guys.

My interests—science, science fiction, computers, technology, math—share something that's not offered by creative outlets such as art and literature or by descriptive studies such as history and some social sciences. I have a deep need for predictability. Not for things to be boring—just to understand how they came to be. If I hear a crash in my living room, I don't say, "Something must have fallen down," I go look to see what fell, and I feel some distress if I can't match the sound I heard with the state of the living room floor.

So it is with computer software. If my computer program doesn't work, I want to know why, how and what I can do about it. I build detailed models of the inner workings of the machine. Not the electrons in their quantum states, but at a logical level. I believe it is this personal requirement of consistency that makes me a skilled computer programmer. I have a sufficiently deep understanding of the machine at all levels that I can easily catch logical errors, and when I can't see a reason for some anomaly, I know how to proceed. Eventually I find a contradiction between my model and the real world; this discovery leads to a solution to the problem, and I update my model.

I don't think most people, either male or female, have this deep need for consistency and logical thinking, but I have noticed that those who do are more likely to be men than women. It is rare that I find a woman who thinks a lot like me. Although less rare, it is still uncommon that I find a man whose mind works like mine. But I still find myself in social groups consisting of me and ten guys. This frustrates me, and so I attempt to join with other women in social or professional contexts. Since the net is a large part of my life, it is an obvious place to seek such contacts.

The FIST Fiasco

A few years ago I joined an electronic mailing list called FIST, an acronym for Feminism in Science and Technology. It sounded like just the ticket; I thought I would be exposed to other women scientists and engineers, perhaps get to meet them in person and maybe establish some friendships with kindred spirits.

I was dismayed to learn that what many members of this list thought of as feminism and feminist thinking bore very little relation to my ideas. I am fond of the quote, "Feminism is the radical notion that women are people." I find no need to prove that women are different from men, that women's way of thinking is different (or better!) from men's, and so on. But the FIST list was full of messages decrying the sexism of scientific investigation, and not just "regular" sexism, that is, the difficulty women face in getting promotions, raises and other recognition in all fields. No, their messages declared that the scientific method itself was anathema to women.

I stayed on the FIST mailing list for a few months. I grew increasingly dismayed, and increasingly depressed, that I could not fit into the mold my body had given me—every time I read a FIST message I'd look for my penis! Eventually some women on the list proposed that the list should be closed to men. At the time, there was a lively debate going on about the feminist view of science, with a few men arguing what was also my point of view. I became outraged that these women would close themselves off because they could not stand to have their beliefs questioned. I resigned from the list.

Sometimes, it can be appropriate to close a list to unwanted outsiders. It is not immoral to have a small community or even a closed community. "Community" is lost at some size, and "politics" sets in. The Internet is not a community, it's a nation. There are things I want to discuss with my friends and neighbors; parties I want to hold without inviting the entire world. So it is with the net—small community mailing lists that do not grow without bound are not *a priori* a bad thing. There is a place both for small communities and for larger societies where any and all may join.

But I resist strongly the desire to close lists solely based on gender. Currently, the overwhelming majority of Internet users are men. Some people, like the FIST proponents, say that the masculine way is to speak without listening, and that the feminine way is to listen and speak only when needed. This is a generalization, and like all generalizations is true at best in the abstract; in the specific it is often false. Some men are considerate; some women are rude. Some men are quiet; some women are prolific. I disagreed with the FIST list's proposal to close its borders; it seemed like a bad reaction to what I felt was well-placed criticism. In contrast, Usenet, the collection of public newsgroups on many topics, provides lots of places for people to voice dissenting opinions without any fear of being kicked out. But there are different standards in each newsgroup, and you may be "flamed" if you don't fit in.

A Little Usenet Advice

Usenet (also called netnews) is an anarchy and is "run," therefore, by those with the loudest voices, or the most time to spend on an issue. There are no representatives on the net. The only way to have your issue raised or your opinion voiced is to speak out. Thus the net is run by those who are prolific, by those who primarily speak and only secondarily listen. Those who only listen are ignored, and those who rarely speak are drowned out. Some of the latter run away from the net after only a short experience, and only a few learn to participate in the net without fear.

Parts of the net can be pretty nasty places. Some people on the network seem to enjoy flaming other people for their views or their ineptness (real or imagined). These flames can be very blunt and insulting—things someone would never say in a telephone or face-to-face conversation. One can learn to accept such vicious abuse of one's intellect—but this may lead to joining in and returning fire. Far better, I think, is to avoid flaming others and inciting others to flame. The net is large enough for both flamers and those who wish to avoid them. I have been an active participant on Usenet for over a decade and have developed a number of techniques for approaching the net in general and Usenet in specific without getting burned. Here, I'll primarily discuss my experiences on Usenet, where messages and responses are available to be read by anyone anywhere on the net, thus amplifying both the highs and lows of communication.

I have frequently heard the advice, when in a classroom or lecture situation, that if I have a question, odds are someone else in the audience has the same question, so I should definitely ask it. This advice is intended to counteract the inherent shyness most people have in large groups. This advice does *not* hold for the net. It is easier to lose your shyness on the net than in a classroom—it feels intimate, since it's just you and the computer in the privacy of your own home or office. But, really, the lecture hall you are in is huge—much larger than any physical site you've ever been in! It is more likely that, if you have a question, someone else has *already* asked it. The answer will be waiting for you shortly. Indeed, some questions are so frequently asked that they have given rise to the FAQ—most informational, and some social, groups have lists of Frequently Asked Questions with answers. Asking something that's in the FAQ, if there is a FAQ for a group, can get you flamed way out of proportion to your transgression. Being the recipient of this sort of abuse can be very painful, especially since it generally occurs early in a netter's career. This is the nastiness that turns people off completely from the net—or encourages them to themselves *become* a flamer! Yet it can be alleviated by simply waiting for the FAQ to appear, which it will in due time.

Before posting, I always check out a group by reading it for a few weeks. It can be annoying to wait, but the emotional savings can be substantial, especially if a group of hotheads is dominating the group. Sometimes I'm just tapping into the collective wisdom of the net—to find out how to do something or to get advice on some problem, whether it is with a computer issue, or some unrelated technical or recreational issue. Even if I'm not planning on becoming a longterm member of the group, simply by doing a little lurking in advance I can determine if my question is appropriate to the group, if it has already been asked and answered, and what sorts of replies I am likely to receive from my query.

.   .   .

Scuba Tanks, Motorcycles and Denizens of Doom

Some Usenet groups are community-oriented—that is, the group has gathered together to discuss a topic, and the population is more or less stable. Although the group's population may be increasing because of the general population increase on the net, posters tend to stick around for significant periods of time and get to know one another. Other groups are more of a resource, and most participants are drop-ins or transients. A few regulars volunteer, in essence, to run a public service: for example, maintaining FTP-able archives of past postings and FAQs.[1] A terrific example of this type of group is *rec.scuba*. People post reviews of scuba service providers in locations around the globe. These reviews are collected into an FTP archive and made available via the FAQ. The posting population of *rec.scuba* tends to write in a fairly professional fashion, with a high signal-to-noise ratio, that is, the interesting or useful posts (signals) outnumber the boring, repetitive, flaming or just plain useless posts (noise). The questions are well-formed, and the answers are thoughtful. Indeed, I've found that most recreational groups targeted to real-life activities, as opposed to online-based activities, contain serious enthusiasts who enjoy helping new people learn about their activity and enjoy talking about their experiences with related equipment and locales. It may appear otherwise to the dilettante: All the posters seem to have so much experience and skill in the activity that they would not welcome a novice posting. I've generally found, though, that as long as I am polite and my question or comment is well-phrased, I will not be laughed off the group, but instead will receive helpful advice.

Noncontroversial topic-oriented groups, by definition, are less likely than controversial groups to be full of flames. Low-traffic groups will also be nicer. Generally in a low-traffic group fewer people post, fostering a sense of community and lowering the probability of a bad apple. In a small community, would-be flamers may come to respect the opinions of others in the group more and tone down their own messages. If a group is too large, a post is just one in a multitude, which is nearly equivalent to anonymity, and it's easier to be nasty when you're anonymous. In a huge group it's easy to feel that *they* don't care about you so why should *you* care about them? This lack of community reinforces itself, and pretty soon you have a group full of hate-filled messages far beyond reasonable expectations. But this is not a hard-and-fast rule: There *are* nice large groups.The best way to tell is to observe for a few weeks.

For example, *rec.motorcycles* is a high-volume community-oriented group where you will find folks who've been around since before Usenet posting right alongside folks who hadn't even heard of the net a year ago. *Rec.motorcycles* is quite nice, especially considering how the naive outsider might otherwise stereotype its posters. I've been active on and off for several years and am a member of the official *rec.motorcycles* motorcycle club, the Denizens of Doom. I'm not active now because of the volume; the group gets well over two hundred messages a day, and I simply haven't got the time. What I've found, though, is that *rec.motorcycles* is generally full of helpful people whenever you have a technical question about motorcycles and motorcycling. Denizens are tolerant of most frequently asked questions and proffer lots of useful advice in response to questions regardless of how trivial or technical they may be. Yet denizens can have a bite, should you trespass into one of the forbidden zones. There are a number of in-jokes, most having to do with the club, references to old posts or (sometimes apocryphal) stories about real-life gatherings. If you ask an obvious newbie-question about one of these in-jokes, you'll get bitten. The only way to enjoy them is to be patient; hints get dropped regularly that will give you a clue to what the jokes are about.

A lot of the social and recreational (*soc.* and *rec.*) groups have real-life get-togethers. For *rec.motorcycles*, there are organized group rides, and once while visiting California I was able to join local rec.motorcyclists for dinner. Attending even just one such get-together can be very helpful for a better understanding of people's messages. The printed (or displayed) word lacks a lot of expressive power that comes from facial expression and voice tone. Meeting other posters can compensate for those losses and soften what might otherwise seem harsh words. It gives you as reader a voice to hear, as though the poster were reading the post aloud, making it easier to imagine appropriate facial expression and tone of voice. Obviously it's not practical to meet everyone in a global network, but meeting even a few can help make Usenet seem more personal.

Warlording

Some groups on the net have nastiness and flaming almost built in. Certainly such groups as *alt.flame* are solely dedicated to flaming and have raised it to an art form: you get flamed if your flame isn't hot enough. But there are also many political groups you would hope would be a place for informed debate, yet which devolve into name-calling at the slightest provocation. *Talk.politics.abortion* is the canonical example of such a group. People have strong views, and they express those views in the strongest possible terms. To make matters worse, the attitude of "if you can't stand the heat, get out of the kitchen" prevails in many Usenet groups. Quiet, sensitive, thoughtful people are drowned out. I stay away from such places. Even if you too have strong feelings, there is little point in subjecting yourself to this sort of abuse in the vain (and it *is* vain) effort to change the minds of a few people. Far better to support the local real-life grassroots organization that promotes your point of view, or to write your government representatives, agencies or local or national news and educational media to express your views.

Some of the topic-oriented groups discuss an online topic rather than a real-life topic. It is much easier for one of these groups to develop a bad attitude or bad reputation—somehow grounding a topic in real life helps foster an attitude of friendliness or helpfulness.[2] And some of these online topic groups go out of their way to be unfriendly.

Besides *alt.flame*, another such group is *alt.fan.warlord.* The origins of the group's name may be obscure, but the group's purpose is very evident: to publicly ridicule a certain class of net newbie and net idiot—those with preposterous .signature files. Being "warlorded" is a not a pleasant experience.

Sigs

A .signature file automatically includes your contact information, email address, and so on into a posting, saving you the trouble of remembering to sign a message and preventing you from making embarrassing typos in your email address or other pertinent information. Signature files can be quite useful, and thus the feature is included in most posting programs. However, it opens up a whole world of potential silliness and abuse: Arbitrary text can be included in all of one's postings with nearly zero effort. When something is effortless to do, one doesn't necessarily take into account the impact one's actions have upon others. And when something is automatic, as is the inclusion of .signature files, one doesn't necessarily realize or remember that this is occurring, or even what the .signature file contains. This can have embarrassing results, for instance, when someone has edited their .signature to contain a joke and forgets to remove the joke while posting on a serious newsgroup, or if someone overreacts to the ability to have a .signature and creates a large, complex file full of difficult-to-decipher, allegedly artistically presented information, which keeps getting posted over and over, creating an eyesore for readers across the globe.

Let me include my own .signature file and explain why it collects compliments rather than flames. First, I have several signature files, all very similar, but slightly tailored to the group I'm posting to—I recommend this approach, as it permits you to make in-jokes for one group without forcing them upon another group that may not care or may deride you for it. Since I have more than one, I must choose which to use, and manually insert it, raising the effort level and thus the amount of thought required. Sure, computers are supposed to reduce effort, but they should not reduce effort so far as to eliminate thought! This is my *rec.motorcycles* signature:

Judy Anderson yclept yduJ   'yduJ' rhymes with 'fudge'
     yduJ@cs.stanford.edu   DoD #166
Join the League for Programming Freedom, lpf@uunet.uu.net

The good parts: It's very short, and it says clearly who I am and how to contact me. It has one cutesy thing ("yduJ rhymes with fudge"); I've been using this since I began posting on the net, and it has become a sort of trademark. Thus it lends consistency; people recognize me from one group to another by this phrase. It has one private reference, to my Denizens of Doom membership number, which is good as long as I confine myself to posts to *rec.motorcycles* and related groups with this signature; otherwise people don't have a clue what it means.

The bad parts: It contains an advertisement. The advertisement is short, though, which does alleviate it somewhat, and the League for Programming Freedom happens to be neutral or politically correct within most Usenet communities, so this hasn't gotten me flamed. It contains one archaic word (yclept).[5] This generates primarily queries rather than flames, but still it is the sort of thing that a good signature writer will think about carefully before including.

There are a lot of things this signature does not have: It does not have ascii art. It does not attempt to emphasize itself with a surrounding box of asterisks. It is written in mixed case. There are no spelling errors. It does not have a quote from some well-known personality, or worse yet, a quote from an obscure personality (a quote can be all right if it's short and sweet and useful as a trademark, but it's important to avoid long quotes. Allegedly funny quotes can be problematic as people's senses of humor are quite varied.).

The aforementioned *alt.fan.warlord* group is a reaction to the over-the-top signature, a network-wide blight afflicting all groups. But, as in many such things, warlorders can themselves go overboard, no holds barred and pulling no punches. It is not a newsgroup for the faint of heart, though when in the right mood I get a certain smug satisfaction from being a member of this "elite," laughing at the plebes with their pathetic attempts to be seen as different among the masses. But one definitely has to be in the right mood for it.

Brush Strokes in ASCII

A signature element participants in *alt.fan.warlord* deride with vehemence is ascii art. Ascii art (really ASCII, American Standard Code for Information Interchange, the internal code of most computers which translates into ordinary text) is an art form in which the characters on the screen form the lines that make up the picture. For instance, one can make large "fonts" out of horizontal, vertical and diagonal lines from some punctuation characters. A lot of ascii art is schlock, but a talented and patient artist can do quite a lot with the medium. As one might expect, there is a newsgroup dedicated to ascii art. I have been a frequent contributor, not as an artist, but as a collector. I have been collecting ascii art since I saw my first picture, and my collection is now quite extensive. Even with more than fifteen years of collecting, I am constantly amazed at the variety and quality of art that can be created. As just one example, I show here a piece created by Rowan Crawford, a very prolific ascii artist from Australia.

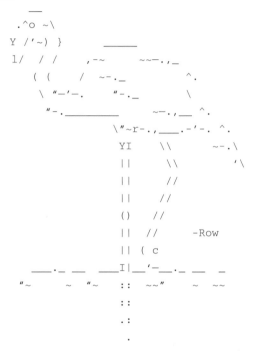

The choice of just a few individual letters, punctuation and spacing (which make up the brush strokes) can make the difference between a rough, ugly picture and an aesthetic one. Of course, more of the former gets posted, because not everyone is as talented as Rowan.

It is every *alt.* group's dream to upgrade to a mainstream group.[4] I helped in *alt.ascii-art*'s attainment of that goal: There is now also *rec.arts.ascii*, with primarily the same content, though a slightly different population and tone. To upgrade, one has to go through a fairly lengthy process of writing a proposal, participating in a discussion period and, finally, waiting out the voting period. I felt I had two jobs in this campaign: First, to help make sure the proposal was acceptable to the Usenet cabal, and second, to prevent young, bubbly *alt.ascii-artists* from inadvertently biasing the vote against the group by bringing their misplaced enthusiasm to older, more staid groups during the campaign. There isn't really a Usenet cabal, conspiracy theorists aside; it is simply that some people prefer to spend their recreational network time by commenting on all new group proposals. A lot of these people are smart and experienced and know what does and doesn't work in Usenet groups. Another group is made up of hotheads who have strong opinions, whether right or wrong, on how the net should be run. It's not all that difficult to accept the advice of the experienced folks while dodging the flames of the hotheads, but it entails some careful wording of messages, and my experience at deflecting flames was helpful in this task.

Alt.ascii-art is a good beginner's group (for those interested in the subject). Where many people go wrong, though, is in bringing the lessons of the "good beginner's group" to the rest of the net. Ascii art gets its bad reputation because some of its fans tend to lose sight of the fact that it is inappropriate to post a three-screenful picture of a vaguely helical shape to *sci.med* on the grounds that it represents a strand of DNA.

The community of *alt.ascii-art* is very much the antithesis of *alt.fan.warlord*. It is full of young, (usually overly) enthusiastic, would-be artists, who are prolific posters of primarily schlock (though, for a collector, the gems in there are definitely worth waiting for). They bubble about their own cool art and other people's cool art, they ask huge numbers of Frequently Asked Questions, they can't spell and they repost messages multiple times per week. In contrast, *alt.fan.warlord* members usually have several years' experience under their belt and have seen it all, over and over and over, so they make fun of everyone, even those whose transgressions are quite small. You might think it difficult to be a member of both communities, but by altering my frame of mind appropriately I find it possible to view the same .signature as beautiful on *alt.ascii-art* and laughably inappropriate when quoted on *alt.fan.warlord*.

What Really Counts

The phrase "On the Internet, no one knows you're a dog" appeared in *The New Yorker* in a P. Steiner cartoon in 1993.

It is something that one should take to heart. Usenet, while it can be nasty, acerbic, uncaring and unsympathetic, is a truly nondiscriminatory society. It judges you only through your postings, not by what you look like, your marital status, whether you have a disability, or any of the other things that are traditionally used for discrimination. I have read *misc.handicap* postings written by a deaf-blind woman; nothing in the postings showed any sign of disability. If you have a cooperative administration, even your name (and thus presumed gender) can be masked or changed.

In general, it is your opinions and your manner of expression that really count. I like to think that people respect me; the responses I receive to my postings confirm that they do. My postings are not immune to being flamed, but usually I am expecting it, whether it is because of the opinion I express or the manner in which I express it. I feel I have a lot of control over how people perceive me. What works in one group may not work in another, but with careful reading, I've determined what will be accepted, appreciated and even admired. By paying attention to simple things such as using proper spelling and adequate grammar, as well as more subtle things such as the mood and traditions of the target group, you can become a well-respected poster, whether it be as an asker of intelligent questions or a good source of information.

Endnotes

1. FTP refers to File Transfer Protocol, a standard Internet protocol that allows a user to log on to a remote computer and transfer public files to their own computer.

2. A big exception to this generalization is *talk.politics.abortion*, which is completely flame-ridden.

3. *yclept* is the past participle of *clepe*, an archaic word meaning "to name or call."

4. Mainstream groups are *rec* (recreational), *soc* (social), *talk* (discussion), *sci* (scientific), *comp* (computer related), *news* (relating to Usenet itself) and *misc* (all other categories). *Alt* is a separate hierarchy with different politics and rules for group creation and propagation.

MALE AND FEMALE

THE NET CREATED THEM

The Price of Admission: Harassment and Free Speech in the Wild, Wild West

Stephanie Brail

Online harassment has become a media headliner in the last few years. I should know: I was the target of one of the more sensationalized cases of "sexual harassment" on the Internet. When I wrote about my and others' experiences with online harassment, I found myself inundated by requests for interviews with other reporters writing the same story. I've been quoted in *USA Today*, interviewed by *Glamour*, pursued by the local ABC news affiliate and pounced on by editors at *Mademoiselle*, who wanted, I assume, a juicy tale of cyberspace stalking to sell more issues of their magazine.

Online harassment is a tough issue. Finding the fine line between censorship and safety and creating a better environment for women in cyberspace, are complex tasks. As I've wrestled with these issues, one of the sharpest areas of concern for me has become the effect harassment has on our most precious online commodity: Free speech.

Wanna Fuck?

Just what is online harassment? If someone sends you a request for sex in email, is that harassment? What if someone calls you a name online? A woman is called a "curmudgeon" and complains that the poster is harassing and slandering her. Is he? Many might define online harassment as unwanted, threatening or offensive email, "instant messages" ("sends" or "chats" on some systems) or other personal communication that persists in spite of requests that it stop. But this is a poor definition, because what is unwanted, threatening or offensive to one person may not be to another. Sometimes it seems as if the definitions boils down to a personal one of "I know it when I see it."

There is a huge gap between legal definitions of harassment and what we describe as online harassment in common parlance. The legal aspects will come later; for now, let's look at the nonlegal definition.

Much of what is termed "online harassment" is "wanna fuck" email. A "wanna fuck" is simply an email request for a date or sex. An email asking for a date is not in and of itself harassment, but what bothers many women on the Internet and on online services is the frequency and persistence of these kinds of messages. America Online's (AOL) chat rooms, for example, are notorious for having a barlike atmosphere...should you enter a chat room using a woman's login name, you're likely to find yourself the target of a wanna fuck "instant message" from some man you've never even heard of. Though AOL has strict rules of conduct, called their Terms of Service, which explicitly ban harassment, as well as obscenity, chain letters and other "offensive" types of online communication, the staff at AOL is hard-pressed to be at all places at once, so the Terms of Service do not guarantee a "safe environment," however hard AOL tries.

I would guess that wanna fuck email generates the bulk of online harassment complaints, and that repeated, targeted harassment of the kind I experienced is actually quite rare. So I'm not sure if online harassment has become a media hot-button because it is a matter of concern or because it creates another sensational headline, or both. So much hype and angst has been whipped up over this issue, it's hard to look at it objectively anymore. Many users get really riled up about having a safe environment online, but equal numbers, many of them women, are so sick of this subject they don't ever want to see another article about it as long as they live.

Enter the Online Harassment Poster Queen

That I had a harrowing online experience in 1993 was one thing. That it brought me my fifteen minutes of fame was more disappointing, to say the least. When I first spoke out about online harassment, I meant it as a call to arms, a message to women that it was time to take hold of the keyboard and carve out some female space in the online world. I was apparently riding a wave of media interest, set in motion by some genuine activism on the part of many dedicated women activists and computer mavens, but the bigger force was, of course, the American lust for a new victim.

My experience of harassment coincided with an article I was writing on sexism online for *On the Issues,* a small feminist quarterly out of New York. When my editor found out I was being harassed, she thought it would be great to add that personal touch to my story, which turned into an article about online harassment called "Take Back the Net."

When I began writing the article I noticed that what seemed like every other reporter and freelance writer in the business was working on the same story I was. Next thing I knew, I was being interviewed about my experience instead of writing about it. That was 1993, when the Internet was still just a blip on the national media scene. If the number of interviews I did is any indication, harassment took up an inordinate amount of ink that year. Two years later I was still getting calls from reporters.

I believe these stories of online harassment are told and retold partially because of the "car wreck fascination" factor, but more importantly because we all keenly feel our vulnerability in the new medium of computer-mediated communication. Women, especially, need to discuss and understand the implications of online harassment because it affects our ability to use the medium and, thereby, to take part in something that will only become more important to our freedom. How many women would have voted had polling places been in dark alleys?

The Online Car Wreck

Here's my story. It has become, even in my mind, more of a sound bite than something real. The gory details have been swept away in the interest of something quotable; the actual event a faint memory while the intellectualizations I created around it abound.

What happened is less interesting than why it happened. I was harassed not because I was an innocent bystander, or another female using the Internet, but because I had a mouth. I dared to speak out in the common space of the Internet, Usenet.

Usenet is a collection of online discussion conferences or forums available to almost all Internet users. My boyfriend and I had been reading the Usenet newsgroup *alt.zines* to discuss underground, homemade publications, because we were in the process of creating our own zine.

I don't remember exactly how the flame war/argument started, but a young woman had posted to the group a request to talk about Riot Grrls zines. Riot Grrls is a political and social movement of young punk postfeminists, inspired by girl bands like Bikini Kill and the Breeders, and a hallmark of the movement is the numerous fanzines created to support these bands. At the mention of Riot Grrls, some of the men on the group started posting vehemently in protest. They didn't want to talk about those stupid girl bands; the girls couldn't play anyway. Someone suggested that the young woman start her own newsgroup called "alt.grrl.dumbcunts."

In spite of having been online for years, I had never really participated in Usenet before and had no idea how much anti-female sentiment was running, seemingly unchecked, on many Usenet forums. When I saw the treatment this woman was getting in response to her request to discuss Riot Grrls, I was not only appalled, but also incredibly angry.

The woman who wrote the original note fought back, posting angry, curt responses to the one or two men who were leading the charge against the "stupid" Riot Grrls. My blood pressure increasing, my heart pounding and my body aching for justice, I joined the fray. I'm a natural writer—wordy, passionate—and, in a world where you are your words, I am loud. I bellow, I scream, I prognosticate. I was writing what I thought at the time were noble words, defending the honor of all women.

That was my first flame war. Probably my best. What an ego-driven experience! I had fans of both sexes emailing me letters of encouragement. Most of my detractors responded with a lot of sexist drivel, and several people, who identified themselves as Internet old-timers, tried to explain (to my deaf ears) that I obviously must be a newbie or I wouldn't be getting so upset. (Looking back, they were right; being online for a while makes you increasingly blasé about online slights.) To a certain extent, the whole thing embarrasses me now, but at the time, I didn't think I was doing anything wrong. I felt I had to speak up, largely because a few men were telling us women to sit down, shut up and go away.

It's hard to explain the kind of high you can get while participating in a flame war; in some ways it's like being on a roller coaster—your stomach may be churning, but it is a delicious kind of sickening feeling, steeped with adrenaline. I had never participated so much online. I came home from work dying to see what the responses were to my posts. Then the harassment started.

I was not the first target. One of the women sticking up for Riot Grrls, perhaps the one who originally started the topic, received obscene email from a guy named "Mike." The email was anonymous—sent with no real name and with a fake return email address. She posted the letter to the group to show how the flame war had degraded. Others received similar emails. Then my boyfriend, who had been one of the guys sticking up for women, received a few nasty ones, asking him why he supported Riot Grrls—"fuck 'em, their daddies did," one anonymous email said. Another one said: "Heh heh—I'd love to see a porno with a father doing his Riot Grrl daughter—she has a bad haircut and is wearing boots with a pink mini. He says, this will give you something to rant about! as he sodomizes her little riot ass."

"What should I do with this?" my boyfriend asked.

"Just ignore it," I responded. "What a jerk."

Easier said than done. My boyfriend posted one of the notes back to the group anyway, with a sarcastic message of disapproval. Even though "Mike" had no idea my boyfriend and I knew each other, soon after that I became the target.

When I received email with the word "cunt" splashed across the screen, I became sick to my stomach. The harassment was a shock; in spite of the mess on the newsgroup, I hadn't expected it. But I was shaking, less from fear than from anger. I tried to email a response to the guy, but the message automatically bounced back to my mailbox, compounding the insult. So, as everyone else had done, I posted the note back to the group, coupled with some very nasty comments.

In response, I found more email messages in my box the next day, and the next day, and the next. Reams of pornographic text detailing gang rapes. Strange, poorly formatted messages full of long ramblings about how the poster was a writer and how he found all this so interesting. There were details about a girlfriend, Valerie, who purportedly worked at some great book publisher in New York. He was harassing me because he was going to write a story about it, he told me.

Each message was from a different fake email address, with a different name on it. I had no way of telling which messages were from friends and which were from my foe. It made me sick to read much of the stuff he sent me, but I went through most of it, trying to find a clue as to who this person was.

At this same time, a man on the group sympathetic to our side, Ron, was receiving several emails a day from the same person, although his were less frequent and much tamer. Ron had taken up my cause like the proverbial knight in shining armor. I didn't know him but was relieved to have an ally. We were now battling detractors on the newsgroup, who were sick of the flame wars and totally unsympathetic when we posted public notes telling Mike to stop. Some even told us that by complaining about Mike we were censoring him!

Mike wreaked havoc with my email inbox for several weeks. He wrote a story about the incident, which he posted to the group, but in his warped version of events, I was supposedly turned on by the whole thing. He also faked some posts to another Usenet group, to make it look as if I had posted something he had written. It was only when I got a strange message from someone from *alt.sex.bondage* that I found out Mike had been emailing people there and putting my name and return address on the messages.

Ron and I tried to get help from the system administrators at the university from which the posts originated, but to no avail. The sysadmins told us the only way they could catch him was "in the act." We considered calling the police and the FBI, and only after we made the threat "in public" on the newsgroup did Mike's email slow down to a trickle, though I continued to receive occasional pornographic email from him. Months later I received an email from him at one of my other email addresses. I have no idea how he found the address, but the message he chose to send was chilling: "I know you're in Los Angeles," he wrote, "Maybe I can come for a date and fix your 'plumbing.'"

By this time I was incredibly paranoid. I made sure the doors to our bungalow were always locked; I practiced self-defense. When a male friend called us and left a prank message, I thought Mike had found our number, and I panicked.

But finally, Mike goofed. He sent a message to my boyfriend that left some tracks. My months of dealing with the inner workings of Internet mail paid off, and I was able to track him down. I forwarded the message to his real email box without comment, and I haven't heard from him since.

Although the experience was horrific, it was a tempering kind of fire. It forced me to learn UNIX, the computer language much of the Internet is based on. In response to similar types of harassment, other women created their own spaces to be free of attacks of this kind. In response to the events on *alt.zines*, one young woman began a private female-only mailing list called Riot Grrls, which is still going strong. The support and advice I received on that mailing list during that time were invaluable in keeping me sane and active on the net.

The whole incident has taken its toll, though. I don't trust that this is the last I'll ever hear from Mike, or anyone else, for that matter. I'm careful what kind of information I give out online now—never my home phone number and certainly not my home address. I certainly know how easy it is to make an enemy on the Internet, and I stopped participating in *alt.zines* long ago. I'll probably never post there again. And that's the true fallout: I've censored myself out of fear.

The Big C-word: Censorship

I've censored myself. My choice, right? I'm not so sure. Do I or do I not have the right to speak my mind in public without being harassed, stalked and threatened because of what I say?

The Internet is the Wild Wild West—as far from the civilized, or at least patrolled, corridors of the commercial online services such as Prodigy as the West was from the streets of Boston. And just as it's easy to romanticize the Wild West, forgetting the abuses that took place during that savage time, it's easy to romanticize these pioneer days of the Internet as well. I myself have loved this time of openness on the net, when relative freedom and a lack of government control made it one of the coolest places to be. It saddens me that some people abuse the freedoms many have taken for granted on the Internet, and that these freedoms are now threatened thanks to such immaturity.

It seems that a truly free space for public discourse is too threatening to the American public and we've only begun to see the start of what's likely to be a long and drawn-out fight to keep alive the delicious anarchy that's been such a fertile ground. Without free speech, the Internet will be as lifeless as, well, corporate broadcasting.

At the same time, I believe that online harassment is, to some extent, already killing free speech on the Internet, in particular the free speech of women, although women aren't the only targets of these vigilante censors.

Shut Up or Put Up

Unfortunately, because of rightful fear of government control, many people see this harassment issue as one that shouldn't be mentioned. And many don't believe it should matter anyway. Just fight back, they tell you. This is easy advice for a loud-mouthed, college-aged know-it-all who has all the time in the world, but does it apply to real, working women, who don't have the time and luxury to "fight back" against online jerks? And should we have to, as the price of admission? Men don't usually have to jump through a hoop of sexual innuendo and anti-feminist backlash simply to participate. They use their energy for posting, while we often use ours wondering if we'll be punished for opening our mouths. And with all our training to be "nice," are most women even prepared to do such battle?

This is not to say that supportive people aren't out there to help. When I was being harassed, new Internet friends from all over the world offered me technical assistance.[1] Many gave freely of their time and knowledge, and some offered to help me construct mail filters to keep out the offensive messages. Some offered to track the harasser down. Many friends offered to email bomb the perpetrator in return, but I declined.

For years a laissez-faire attitude has governed behavior on the Internet. Users didn't turn to lawsuits to solve their problems; they dealt with them using the technical tools available. Any talk of regulation scares users. When I first started talking about online harassment, people criticized me for trying to bring the regulators down on our heads. They should have been yelling at the jerks who abuse the system. For speaking out against online harassment I was likened to an Andrea Dworkin disciple, or worse, Phyllis Schlafly, out to wipe the Internet clean of smut.

Pornography Is Not Harassment

Despite those who believe that certain types of sexual content are harmful to women, there is a difference between pornography available online and harassment. If someone wants to post nude pictures to a newsgroup, I don't have to see them. Not only would I have to decide to go to the newsgroup myself to see the pictures, I would have to download them *and* decode them *and* have the proper configuration on my computer to see them. If I accidentally went to one of these newsgroups, all I would see on my screen is a bunch of garbled text: the encoded version of the smut. (And maybe some lewd words, but that's about it.)

Unwanted erotica and pornography do become more of an issue with the World Wide Web, where the availability of embedded graphics makes it harder to avoid the online equivalent of *Hustler*. A friend looking for domestic violence resources on the net checked out an address and within two "clicks" was looking at pornography. A page linked to another page linked to another page, and she'd gone from an informational Web site to a pornographic one. It can be disconcerting.

Fortunately, new software is making it easy to avoid explicit material on the Internet and World Wide Web. All you have to do is screen it out. But when someone starts sending detailed descriptions of gang rape to my email box as a veiled threat or starts to post pornographic stories with me as a main character, the issue has gone far beyond pornography. I am concerned that harassment and pornography have somehow become confused in the minds of our lawmakers: The harassment issue has been co-opted to create an excuse for banning so-called indecent material. The two are not the same at all.

So What About the Law?

From my discussions with many women online, I have found that most forms of online harassment are mere annoyances, desperate men looking for sex in the electronic ether and hitting on anything vaguely female. To give them the benefit of the doubt, many do stop when asked; many don't mean to hurt people. Many (and I've heard from some of them) are really not trying to scare anyone and are simply trying to make new friends. They may be kind of awkward and clueless, but they're mostly harmless.

Women are often annoyed and put out by this behavior, but as many of the strong women online will tell you, they can handle it. The problem is when the date requests (or "wanna fucks") continue after you've said no twice, or when you're sent repeated email messages calling you a "bitch" for stating something on *alt.feminism.* The question then becomes, how does this atmosphere affect the culture, and does it discourage women from being online in the first place? Is this behavior against the law? And can women speak in this atmosphere?

When I was harassed, I thought a lot about going to the police, but I didn't relish being the start of a high-profile online harassment case, and at the time I thought it would be incredibly difficult to prosecute.

According to Mike Godwin, staff counsel for the Electronic Frontier Foundation, the legal definition of harassment does not apply to most online harassment cases, since harassment is something that technically occurs in a school or work environment. However, civil and criminal laws that deal with issues of online harassment do exist. For example, on the civil side, you might sue for "intentional infliction of emotional distress." It is also against the law to misappropriate someone's name or license, that is, to send mail under another person's name. And Godwin points out that laws of defamation and libel also apply to the online world. In addition, federal laws exist that outlaw threats "through a means of interstate commerce."

Unfortunately, these legal remedies are often either unknown or misunderstood. And, while the good news is that someone can't threaten you in email legally, not every wanna fuck email is legally a "threat" (which is probably for the better in the long run, what with lawsuits running rampant in this country). As Godwin says: "It's not whether you feel threatened, it's whether an objective person looking at it would say it was a threat."

It's also perfectly legal to insult someone in public or by email. While sometimes I believe the laissez-faire attitude regarding net behavior goes overboard, those who call for banning words that hurt are way off base. When I see women call for strict email conduct rules (as I've seen on the women's online service, Women's Wire), or when women call it sexist and harassment when their Web page are linked to the "Babes of the Web" site, I'm concerned that their fear is constricting free speech as much as real and perceived harassment might be.

Dealing with the Current Atmosphere

Because not everyone is going to file a lawsuit, and not everything is prosecutable, women have come up with many different ways to battle the bombardment online. Many women I've talked to have resorted to using male or gender-neutral names to avoid getting hit on online.

One young woman I spoke with, a college senior majoring in English, decided to put "MRS!" next to her name on all her electronic correspondence because of the constant requests for dates. Her comments on sexism online included:

> I think (the Internet) is the last bastion of real ugly sexism because it's unmoderated and faceless. I've received more 'wanna fucks'... and 'shut up bitch' mail than I care to count. I've posted to *alt.feminism* and had men posting me back screamingly hateful email calling me everything from a lesbian to a whore. One man told me that as a woman 'you have so little to complain about in

real life that you stay on the net all day whining about how bad things are.

I've talked with system administrators who've dealt with harassment on the Internet Relay Chat (IRC) simply by shutting down the service altogether. The IRC is a CB-like collection of live chat "channels," much like the notorious chat rooms on America Online. Like those on America Online, many IRC channels are harmless, fun places to hang out, but many others are places where certain men apparently like to "camp out," waiting for an unsuspecting female to log on. It is unfortunate that shutting down the system has been one of the only ways to deal with annoying people online.

While it can be said that "wanna fuck" email is "only words" and "not real," I can't help but wonder how many women are discouraged from speaking up online for fear of being targeted for some sort of sexual advance or another. I wonder how many women have stopped posting their words because they were sick of constantly being attacked for their opinions. I'll be the first person to stand up for good old-fashioned disagreement and even flaming, but I have a problem with women being silenced through sexist attacks and vague physical threats. It is the threat of the physical behind the virtual that makes online harassment a very scary thing.

Sandy's Story

Sandy,[2] a polite and friendly forty-year-old woman with a soft Southern accent, loves cats and frequented the newsgroup *rec.pets.cats.*

In 1993 a gang of people from several newsgroups, *alt.tasteless, alt.syntax.tactical* and *alt.bigfoot,* "invaded" the *rec.pets.cats* newsgroup. By the time the invasion had ended, Sandy had received death threats, hate mail and harassing phone calls, was having her email monitored at work and had almost lost her job.

The incident began when one of the invaders who joined her newsgroup posted a message asking if he could get help destroying his girlfriend's cat. He said the cat was bothering him, but he didn't want the girlfriend to find out if he killed it. When he began discussing poison and drowning as options, Sandy spoke up.

First she sent email urging him not to kill the cat, but if he insisted, to have it "put to sleep" humanely. When the email didn't help, Sandy became concerned, then terrified for the cat. She had nightmares. Eventually she wrote a letter to the police that was subsequently distributed on the Internet.

The flame war exploded. The request for help in killing the cat was actually a fake. The poster and his friends had purposefully chosen a quiet little newsgroup to start a flame war of mythic proportions. Their stated goal was to inflame the members of the group with their posts. And it worked. But when Sandy contacted the police, the invaders became ugly and turned their attention to her.

Soon Sandy found herself on the member list of a *Net.Invaderz* FAQ (Frequently Asked Questions document) that was being passed around Usenet and even several computer conventions. Rather than being a victim, Sandy was singled out as one of the victimizers. "Those of us that opposed the group coming in and invading us (were added to the list)," she said. "It was spammed all over the network as a true document with our names on it."

Sandy was disturbed but tried to ignore the problems as much as possible until she found herself under investigation by her own company. An irate "U.S. taxpayer" had written her employer complaining that he didn't want the Internet used for actions such as those described in the *Net.Invaderz* document. "I'm a twenty-two year employee with this company, with a good reputation which is now in the pooper because of this," she said.

Sandy hasn't prosecuted but the incident exhausted her and made her fearful. She no longer participates in or even reads *rec.pets.cats*; concerned friends email her posts of interest privately. She cannot afford her own home computer, so she can only access the Internet through work, where her supervisor now watches her every move.

Because she acted (in this case alerting the authorities to what she believed to be cruelty to animals), Sandy became the target of a vicious attack launched by a group of people she had never even met.

In part, the wars going on in cyberspace are cultural wars. Who is to decide what is polite and acceptable? Some time ago, I talked with one of the founders of *alt.syntax.tactical*, who calls himself Antebi. His response to those who suggest his tactics are uncivilized? "Learn to use killfiles," he says. "Grow up, welcome to reality."

After talking with him, I understood his group to be somewhat like an Internet fraternity, a bunch of young men who like to do virtual "panty raids" on unsuspecting newsgroups. They per se aren't the problem (I do not think *alt.syntax.tactical* was responsible for the death threats to Sandy), but that kind of mischievous mentality, coupled with a lot of free time, means that certain people can abuse their power in the virtual world.

But should the virtual world be one where war is the only metaphor? An invading army swept through Sandy's village, and when she reached out to protect someone else, they turned their sights on her. She was attacked, accused, harassed and threatened—with no possible recourse. The army captain merely says she should have armed herself. But perhaps there are other ways to live than by the rule of the strongest? Isn't that what civilization is supposed to be about?

Tools, Not Rules

A popular phrase you'll hear on the venerable California-based online service, the WELL, is "Tools, Not Rules." In other words, don't regulate the Internet, train people how to use it and let them decide for themselves what they want to read and see.

I'm all for it, since I believe that overregulation would stifle the Internet. Women can and should learn more about their online environment so they can exert more control over their corner of cyberspace. The move of many women to create mailing lists and online services is a positive one. Rather than playing the victim, we can take charge and fight back with the same tools being used against us.

But the Tools, Not Rules philosophy has its limits. On the WELL, a small cybercommunity of 12,000, where such issues of free speech and community are cherished and routinely thrashed about, user Preston Stern wrote:

Like any other good thing, though, embraced wholly with no conditional moderation [Tools, Not Rules] can easily be turned over and create effects opposite to those intended.... We can insure that everyone has equal access to the tools, but we cannot guarantee that everyone will have equal proficiency. This means that some people, by virtue of having more expertise, more time and/or more experience with the tools, are able to become more powerful, to bend the public discourse and agenda toward their own ends.

Stern wasn't writing this in response to a topic about online harassment, but the concern is the same. Tools can empower, but they can also be a barrier. Women, especially, have a greater problem using Internet "tools"—the typical barriers being lack of time and knowledge and the male domination of all things technical in our society.

Whose Responsibility?

Harassment isn't just a women's issue. In this kind of free-for-all climate, the only people who will have free speech are those who have the gall to stand up to threats or frequent requests for sex, and those who have been lucky enough not to step on the wrong person's toes yet. And while women bear the brunt of this climate, men can also be affected. The man who spoke up in my case, Ron, was harassed and at one point challenged to meet his attacker "face-to-face"—for what, we can only imagine.

Is this the atmosphere that encourages enlightened discourse and free speech? Sandy compares the current atmosphere online to the dark science fiction movie *Blade Runner:*

> It's like another world, it's like another planet. It's like a totally unregulated dirty nasty little underworld. It's got some really nice, great, shining pockets of humanity and education and conversation, and then it's got this horrible seamy gutter-ridden filth...they're spreading like a cancer. As far as how to eradicate that without cutting out the good, I don't know what's going to happen to it. I really sincerely do not think censorship and government regulation is the way to go, I just wish people were a little nicer to each other.

So what can be done? Most women will continue to receive wanna fucks, and many will not even prosecute when they do receive a legitimate threat.

I don't think a legal remedy is the real answer anyway.

Like Sandy, most women I know online are opposed to censorship. I would rather put up with the harassment than have Uncle Sam reading all my email. But I don't think that living with harassment should be necessary to enjoy the Internet, nor do I think the current "everything goes" environment is healthy. I think we can take steps to make the online world a little more safe. Part of what I would consider to be healthy would be an environment where community responsibility, not rampant individualism, was more the emphasis.

Unfortunately, whenever you so much as mention that you want something done about harassment, you are accused of being pro-censorship. Certainly, the strict rules you can find on online services such as Prodigy and America Online are double-edged swords. Perhaps these services are a little "safer," but is that truly free speech? Maybe the price of freedom is tolerance. Tolerance of jerks who want to put up a "Rate the Babes Home Page," tolerance of a few unwanted emails, tolerance of women online. But sometimes it feels as if the price of freedom also means I must be willing to risk my personal safety for free speech.

In real life, harassment isn't confused with free speech. If I get death threats through regular mail and I report that to the police, am I "censoring" the person who sent the threat? Threats are not free speech. Extortion is not free speech. Defamation is not free speech. Shouldn't the question be: Do we really have free speech on the Internet in its present form? Isn't the tyranny of vigilante bullies, however rare and arbitrary, the same as tyranny by an officially sanctioned body like a government or corporation? When people tell me the Internet is just words I can't help but remember checking the locks on my house, looking for a young man who might have decided that words weren't enough.

Easy answers are hard to come by, and extreme positions on either side will do more harm than good. An Internet police state, for example, would undoubtedly not have the freedom of women as its first concern.

Although I would hope that our vigilante friends would take responsibility for their actions and realize that each abuse bodes ill for their and our future enjoyment of the Internet, the burden of action lies with ourselves. Women must take action. The more of us that speak up, the more of us that exist online, the harder it will be to silence us.

Perhaps there are places that we won't want to go to—if a place offends us, perhaps we should just stay away—but instead of withdrawing totally from the online world, with all its riches and opportunities, we can form our own networks, online support groups, and places to speak. We can support each other in existing online forums. Women cannot be left behind, and we cannot afford to be intimidated.

Endnotes

1. Thanks to such help, I now know exactly how "Mike" faked his email messages to me, through a loophole in the UNIX mail system that anyone could exploit. (Mike's not a hacker, but a hack.) Remember: There is no guarantee on the Internet that you are talking to whom you think you are.

2. Sandy is a fictitious name. The woman described prefers to remain anonymous.

Sex, Fear and Condescension on Campus: Cybercensorship at Carnegie Mellon

Donna M. Riley

Women like sex, but Carnegie Mellon University doesn't want you to know that. With its online censorship policy, Carnegie Mellon intervenes in virtual sex, supposedly to protect real women, yet does virtually nothing to help real survivors of rape, battering and sexual harassment on campus. CMU wants to shelter the university community from images of sex online to make invisible the reality of both sex and sexual violence on campus. CMU, which has one of the top computer science departments in the country and developed the first campus-wide computer network, is quickly becoming the leading school in online censorship.

In November 1994, CMU (now known as "Censor Me Universally") tried to ban many sex-related Usenet newsgroups in the name of protecting women from images of sexual violence. Campus feminists, smelling hypocrisy and patronization, formed a direct-action group called the

Clitoral Hoods to send the administration one clear message: We're big girls who don't need to be protected from a bunch of horny geek fantasies. We need access to contribute to online dialogues about sex and create our own sex-positive spaces.

The Administration Acts

On November 3, 1994, Computing Services distributed a memo describing a newly instituted policy for Usenet newsgroups at Carnegie Mellon. The policy seemed reasonable as stated:

The only criterion that will be used to withdraw a bulletin board is that the purpose for which it was established or its primary use makes mounting [sic: carrying] it illegal.[1]

Regardless of the modest intent expressed in the memo, the actual decision[2] was to remove the following newsgroups by November 8, 1994:

alt.binaries.pictures.erotica.*
alt.binaries.pictures.tasteless
alt.sex.*
rec.arts.erotica

The scope of the planned action was far beyond anything suggested by the policy. The asterisks indicate all groups in a hierarchy; the *alt.sex* hierarchy, for example, contains more than one hundred different groups, including serious discussion groups like *alt.sex.bondage*, amateur erotica groups like *alt.sex.stories* and humorous groups such as *alt.sex.hello-kitty* and *alt.sex.bestiality.barney*. One banned newsgroup, *alt.sex.NOT*, discusses abstinence and is, ironically, the antithesis of what the university was trying to censor. Several others (for example *alt.sex.fat, alt.sex.motss, alt.sex.wizards, alt.sex.safe*) are support groups where people discuss their questions about sexuality. Posts on the latter groups are rarely sexually explicit, let alone legally obscene, yet the memo identifies a Pennsylvania obscenity law as the reason for the deletion of the newsgroups.[3]

The Campus Responds

The response on campus was immediate: Within a half hour

after the announcement was made via a local newsgroup, an electronic petition flooded the mailboxes of key administrators.[4] CMU's Student Senate, Faculty Senate and Staff Council all passed resolutions condemning the censorship. Letters from the Electronic Frontier Foundation (EFF) and the American Civil Liberties Union (ACLU) suggested that CMU was overreacting and that obscenity liability was not a major concern for a university carrier of Usenet groups. The administration responded to the outcry on November 7 by agreeing to keep all text-based groups and remove only the sexually explicit binaries hierarchies[5] and by forming a "Bboard Committee" to review the decision.[6] The text was allowed to stay because sexually explicit text is less likely than pictures to lead to obscenity prosecution.

On November 9, 1994, a noontime rally brought national media focus to the university censorship issue. With the theme "Freedom in Cyberspace," speakers included Mike Godwin of the EFF and Vic Walczyk of the Pittsburgh ACLU. No women had been invited to speak until a feminist suggested to the student body president that it would look bad if only men were advocating access to sexually explicit electronic material. I received a call from the student body president around midnight the night before the rally asking me to speak and offer a "feminist perspective" against censorship. As one of very few women who are outspoken about women's issues at CMU, I was the obvious (perhaps only available) choice for a token speaker. When introducing me, the student body president said awkwardly, "She's a...feminist..." at which point the crowd began to boo and hiss; he then added, "BUT...aw, c'mon guys..." and tried to explain, apologetically, to the mostly male crowd that I was anti-censorship and that they should stop booing long enough to hear what I had to say.

The Administration as "Protector"

At the rally it became clear that legality was no longer the university's primary motivation for cybercensorship; protecting women was. Erwin Steinberg, the university's vice provost for education, who became the chair of the Bboard Committee, said in a speech at the rally that the newsgroups should be censored because they contained images of "forcible sex with women, children, animals and the like."[7] Of course, lumping women, children, and animals together as objects of male sexuality begs the question: Exactly what are women like? The implication that women are "like" disenfranchised individuals incapable of meaningful consent—and that allowing the university to be our protector would somehow empower us—angered and amused many rally attendees and galvanized the university's feminist community.

This is a nifty administrative trick: Make horny engineers blame feminist women for censorship imposed by a bunch of powerful white men. Divide and conquer where the 70/30 male-female student ratio predisposes the campus to gender hostility. Posts were seen on local newsgroups about "those damn feminists" and political correctness rearing its ugly head. Meanwhile, the National Coalition Against Pornography and other right-wing groups commended CMU for protecting its female students from potentially harmful images on the net.

But Steinberg's speech was only the first strike. At the first meeting of the Bboard Committee,[8] Steinberg passed around a folder containing images of nonconsensual sex, bondage, bestiality and consensual gay anal sex, images that he maintained were downloaded from and representative of the censored groups. Steinberg reminded the committee members multiple times that they didn't have to view the contents of the folder if they did not want to. The graduate student representative, who was a woman, attempted to decline to view the pictures because she knew they would be offensive and felt that legality—not sexuality—should be the issue. She was badgered by the university's lawyer, Jackie Koscelnik, into examining the images, even though the men were allowed to decline without protest.[9]

At that meeting there was no discussion of whether the content was legally obscene, only whether Committee members found it offensive. The point on the agenda associated with the pictures read:

> If posting posters or calendars of scantily clad women is considered "sexual harassment" or creating a "chilly climate," how will a picture of a bound woman being raped by [sic] a ski pole be considered?[10]

Koscelnik, university counsel, was under the impression that CMU could be held liable under Equal Employment Opportunity Council (EEOC) sexual harassment laws for providing sexually explicit material over the Internet.[11] Of course, she did not know how to subscribe and unsubscribe to newsgroups, and had never used FTP. With this gap in basic technical understanding, it is no wonder she didn't understand the difference between a picture of a woman being raped with a ski pole being put up in an office and a binary file being posted to a clearly labeled newsgroup dedicated exclusively to explicit material that had to be downloaded and unencoded before the grainy image could be viewed.

Committee debates inevitably returned to sex, even amid complex technical or legal discussions, and on several occasions Erwin Steinberg was seen toting around the folder with the "dirty" pictures to show the outside experts visiting the committee; he wanted to impress upon them the offensiveness of the material. At a computer conference, Steinberg made the following statement: "On cable in a California city that I visited recently there is a sex channel on which attractive nude young women bring each other to writhing, moaning orgasm by visible digital manipulation."[12] Later, in a discussion with a media legal expert, Steinberg repeated this graphic description and followed with a question on liability that in no way depended upon the specifics of the sexual episode mentioned. How long did the chair watch the cable channel, and was he really doing it out of his responsibility to Carnegie Mellon?

Another concern identified during Bboard Committee meetings was curbing hate speech in order to make the university more welcoming to women, underrepresented minorities and queer students. The university has had difficulty attracting women and members of underrepresented minority groups.[13] Creating a welcoming climate could indeed assist the school in attracting top people in technical fields. However, in reality, a comfortable climate is not a major concern of Carnegie Mellon.

For example, the university has steered clear of other online harassment issues. Scott Safier, a gay staff member who was receiving anti-gay messages every few days for three months in his personal mailbox, was told nothing could be done. No support was offered to assist him in creating a kill file,[14] and no effort was made to stop the harassment at the source, even though the poster's user id and home machine were included in the "from" line of the message header. Safier was told to "just ignore it."[15] The implementation of kill files, arguably the best way to protect a victim of electronic harassment, is not a priority in university-supported mail readers and the use of kill files is not taught in the mandatory Computing Skills Workshop.

CMU also does the bare minimum required by law to make the university comfortable for women. It does nothing to recruit or even attract female students to the campus. It offers no advocacy for victims of rape or sexual harassment, preferring to keep those matters as quiet as possible. When a woman was raped in the park bordering CMU's campus, the university took no action, calling the rape claim "unfounded"; when a man was raped in the same location, they immediately responded by installing prominent lighting. Despite this history, CMU would have us believe it is acting in the best interests of female students and improving the campus climate. Hardly. The only thing the university is doing "for women" is using them as an excuse to censor. CMU administrators would rather protect female students from images, from information, than deal with the reality of sexual violence on campus and the need for CMU to hold actual student perpetrators accountable for their actions.

The Clitoral Hoods

Many feminists on campus feel that censoring newsgroups in a stated effort to protect women from harassment is patronizing to women. Removing the newsgroups robs women of our agency by limiting our ability to confront and discuss issues of pornography and sexuality on the Internet. At the November rally, Mike Godwin of the EFF offered to send my speech to Susie Bright (author and lesbian sex activist) and gave me her email address. I and another woman contacted her to ask for suggestions, and the Clitoral Hoods direct-action group was born.[16]

At our first meeting the Hoods designed a dozen controversial fliers that focused on three main points:

1. The university was condescending to women in censoring certain newsgroups on the pretense that women need to be protected from offensive material.

2. Pornography does not cause violence against women; women need freedom, not restriction, to challenge misrepresentations of female sexuality and explore our own sense of the erotic.

3. Everyone is encouraged to join the fight against censorship.

Some fliers were factual and textual; others were graphic and provocative. The Hoods selected a banned image as a logo—a line-drawing parody of Ariel (Disney's Little Mermaid), naked and clearly enjoying her own sexuality. The more controversial fliers bore slogans like "Oh, Erwin, protect us again!" and "Do good girls think porn is icky? Fuck, no!"

These fliers provoked much discussion, particularly in the School of Computer Science. One female professor tore down a flier because she assumed men had put it up. Some conservative Christian graduate students were offended by the image's suggestion of masturbation. The posters and later discussion brought attention to the fact that the administration was using women as an excuse to censor through their feigned concern about hostile environments and sexual harassment law.

A Last-Ditch Attempt

Erwin Steinberg then invited Frederick Schauer, professor at Harvard's Kennedy School and author of the final report of the infamous Meese Commission on Pornography, to back up his views. However, what Steinberg got from Schauer wasn't quite the conservative backup he'd expected. Schauer thanked Steinberg for being open-minded enough to invite someone who disagreed with him and then told the crowd of about fifty seated in the auditorium of CMU's business school,

> The likelihood of an obscenity prosecution...is, well, pick your term, some would say infinitesimal, others would say minuscule, and others would say really really really really small.[17]

This statement brought hearty applause, cheers and laughter from the audience, made up of mostly anti-censorship students, faculty and staff. The Hoods were beginning to think we had cried wolf, that our posters and leaflets were produced in vain. But Schauer spent the last fifteen minutes of his speech reinforcing the administration's protectionist argument by claiming that women are underrepresented on the Internet and the World Wide Web because the presence of pornography creates a hostile environment. He reiterated his belief that pornography is a cause of violence against women and children, and that speech can often be legally construed as an action.[18]

So the Hoods' sign reading "If speech = action, I wouldn't say 'Fuck You'" was apropos after all, as were the Ariel images we pinned to our chests and backs. The fact that we passed out information about Schauer's involvement with the Meese Commission before the lecture was especially important, given the way in which he attempted to hide his past associations with censors.

After the lecture, the Hoods asked Schauer some hard questions. One of the dozen or so Hoods who attended Schauer's lecture works at CMU on women-in-science issues. To counter Schauer's contention that porn is the source of gender trouble on the net, she cited examples from the current literature which posited possible alternative reasons for the low number of women online. She ended with an impassioned demand for the retention of all newsgroups, saying, "I read *alt.sex.stories*, and I'm proud to admit it."

Feminine Protection on the Net

The Hoods have continued to work on issues of women's sexuality and censorship. Pat Califia and Ellen Willis, two prominent figures in the field, came to speak on campus thanks to the efforts by individual Hoods. The Hoods sponsor a weekly erotic story hour open to anyone who wants to read their own or someone else's works.

Most Hoods continue to be involved in the campus anti-censorship movement; one founded a campus organization called CAFE—the Coalition for Academic Freedom of Expression. This group will see to it that whatever groups are banned will be accessible to the campus community. Currently, all text groups are available, and the Bboard Committee can't decide which pictures groups, if any, should be banned. Ironically, some groups that were never censored are at least as explicit as the censored ones (for example, *alt.binaries.multimedia.erotica*).

The strategy used by CMU is an ancient one: Blame the woman; she spoiled our fun. As usual, women are the objects for discussion, the subject of the images to be banned, the subject of male protection and the object of blame and scorn. The Clitoral Hoods are about turning this strategy on its head, empowering women to protect ourselves with the tools of the Internet: our wits, our keyboards, and our fuchsia flame-throwers. The Clitoral Hoods are about telling Erwin Steinberg that *we* want access to those newsgroups; some of us want to critique the material, and some of us (gasp!) *enjoy* it. The Clitoral Hoods are telling CMU that women matter, that we refuse to be silent objects in this debate and that we are claiming our voice. "Oh, Erwin, protect us again" is more than a mockery of Steinberg's sexual repression and patronizing attitude toward women. It is a challenge, a line drawn in the sand.

Whatever the fate of the sex-related newsgroups, the censorship crisis at CMU succeeded in creating a pro-sex feminist movement on campus. The Clitoral Hoods seized the opportunity to do more than merely react to the administration's censorship; we were proactive in creating sex-positive spaces for women and raising issues around female sexuality. In the end, CMU's campus is indeed a more comfortable place for women, despite the administration's actions; in fact, the university's goals were achieved by the very group it dreaded most.

Endnotes

. . . .

1. http://www.cs.cmu.edu:8001/afs/cs/user/kcf/www/censor/index.html Calling the newsgroups bulletin boards, or "bboards" for short, may be a misnomer, but it is the informal terminology used on the CMU campus.

2. http://www.cs.cmu.edu:8001/afs/cs/user/kcf/www/censor/misc/summary.html

3. Pennsylvania Crimes Code Title 18~5903. Relevant parts of the statute include the all-too-familiar three-part obscenity definition: "(1) the average person applying contemporary community standards would find that the subject matter taken as a whole appeals to the prurient interest; (2) the subject matter depicts or describes in a patently offensive way, sexual conduct of a type described in this section; and (3) the subject matter, taken as a whole, lacks serious literary, artistic, political, educational, or scientific value." "Sexual conduct" as used above is defined as "patently offensive representations or descriptions of ultimate sexual acts, normal or perverted, actual or simulate, including sexual intercourse, anal or oral sodomy and sexual bestiality; and patently offensive representations or descriptions of masturbation, excretory functions,sadomasochistic abuse and lewd exhibition of the genitals."

4. "Petition against New Bboard Policy," on a local CMU newsgroup, *cmu.student.out.opinion,* 3 Nov. 1994.

5. Binaries in this context are files that, when converted with the appropriate software, can be viewed as pictures.

6. http://www.cs.cmu.edu:8001/afs/cs/user/kcf/www/censor/misc/summary.html

7. E. Steinberg, Freedom in Cyberspace Rally, Carnegie Mellon University, Pittsburgh, Pa., 9 Nov. 1994.

8. The Bboard Committee consisted of the student body president, a graduate student representative, the chairs of Faculty Senate and Staff Council, a staff representative, the Vice Chair of the Educational Affairs Committee, and the chair. Additional ex officio members included representatives from Student Affairs, Legal Affairs, the Library, Human Resources, and Computing Services. It is not entirely clear who made the final decisions about committee membership.

9. Information about the Bboard Committee meetings were supplied by four members of the committee. Erwin Steinberg, Vice Provost for Education and the Committee's Chair, published some of the meeting contents; Declan McCullagh, Student Body President, published his meeting summaries on the Internet; and Peter Ashcroft and L. Jean Camp, student representatives to the Committee, related stories to me personally.

10. *official.cmu-news,* 28 Nov. 1994. The ski pole image has been a major focus for debate on campus. What is quite telling is the use of the preposition by rather than the preposition with. Ski poles don't rape people. People rape people.

11. EEOC laws place a responsibility on supervisors to keep a workplace free of sexual harassment. If a supervisor is made aware that one employee is making unwanted advances toward another, for example, the supervisor is required to act to stop the behavior. Simi-

larly, if a sexually explicit image creating a hostile environment were posted in someone's workspace, a supervisor is responsible for seeing that it is taken down. The extension of this law to limit access to explicit material in private or public spaces is analogous to taking *Playboy* out of libraries lest someone take it out and post it where it creates a hostile climate. Again, the focus is misplaced on the material as offensive per se, and ignores the crucial issue of context upon which sexual harassment rests.

12. E. Steinberg, "Living on the Slippery Slope," *Law in the Information Age: The First Amendment, Privacy and Electronic Networks,* Duke Law School, Durham, N.C., 1/26—1/28, 1995.

13. CMU undergraduate population is 69.4 percent male and 30.6 percent female. CMU undergraduate population is 3.8 percent black. Of the female students, 9.6 percent are members of an under-represented minority, and 6.2 percent of the male students are members of an underrepresented minority (that is, Hispanic, black or Native American). Numbers as of September 1994 from the EEO/AA Programs Department.

14. Kill files are read by mail readers to filter out and destroy mail from unwanted addresses.

15. Scott Safier, personal communication, 17 Feb. 1995.

16. The name "Clitoral Hoods" was contributed by a queer man in Ohio, a close friend of one of the founding members. He had thought for some time that it would be a great name for a women's radical direct-action group, and the name seemed to fit our purposes nicely.

17. F. Schauer, speech at Carnegie Mellon, 24 Feb. 1995.

18. This argument was especially weak, considering the source. As an example of speech that is an action, Schauer used a phone conversation and verbal agreement to fix prices between auto industry CEOs. What he failed to mention is that the conversation is not illegal unless the prices are actually fixed (that is, the action has taken place).

19. A flame is a scathing yet sometimes humorous response in a heated electronic discussion.

Cocktails and Thumbtacks in the Old West: What Would Emily Post Say?

Laurel A. Sutton

login: sutton
password: ******
% You have new mail.

No kidding. Sixty-four new messages: a bunch of academic stuff from the LINGUIST list, some conversation on several of the mailing lists I subscribe to (including the Spelling Follies, which gives me weekly plot summaries of *Beverly Hills 90210* and *Melrose Place*, freeing me from my TV) and, thankfully, some personal email from people I love.

But before I can get to work, I just have to check the news and see what the cocktail chatter is today...

% trn Checking for new newsgroups...
Subscribe? Yy/Nn

No thank you!

(Ignoring alt.journalism.print)
(Ignoring comp.binaries.ms-windows)
(Ignoring alt.culture.sardines)
(Ignoring alt.comp.tandem-users)
(Ignoring alt.comp.tandem-users.sigs)
(Ignoring alt.sex.brothels)
(Ignoring alt.sex.fetish.jello)
(Ignoring alt.sex.fetish.orientals)
(Ignoring alt.sex.guns)
(Ignoring alt.sex.pantyhose)
(Ignoring alt.sports.hockey.nhl.det-redwings)
(Ignoring bit.sci.purposive-behavior)
(Ignoring alt.music.pet-shop-boys)
(Ignoring alt.psychology.mistake-theory)
(Ignoring alt.med.phys-assts)
(Ignoring alt.comedy.improvisation)
(Ignoring alt.magick.moderated)
(Ignoring alt.music.mazzy-star)
(Ignoring alt.foot.fat-free)

Oh god, these new groups get weirder by the minute. *Alt.culture.sardines?* What the hell is that? And *alt.sex.brothels?* I can just imagine the kind of traffic on *that* group. Look at this, *alt.psychology.mistake-theory*—now there's a specialty group. And shouldn't that last group be *food.fat-free?* Unless they really do mean "foot"...let's think about something else.

I find myself checking the newsgroups almost every time I sign on, often more than once a day. I don't contribute much myself, but I do follow approximately twenty to thirty different conversations every day. As I sit there scrolling, scrolling, scrolling, I am intensely aware of what a massive waste of time this can be. But I do it anyway. I can't help it; I don't want to miss anything, even though I don't know what it is I might miss. It's a twenty-four-hour-a-day cocktail party, and everyone is invited.

So why do I keep wasting my time like this? The most compelling reason is the simplest one: Human contact. There are those who rant and rave about the impersonality of computer-mediated communication and how it's going to destroy the fabric of our society (a complaint voiced about every new technology since the invention of the horse), but I feel it's just the opposite. Communication through computers can bring people together, regardless of their distance or location. For once in your life, you can be sure that people are paying attention to what you're saying, and not the way you look or how nicely you dress. You contribute to a newsgroup from your quiet little room (or big, fluorescent-lighted work space) and within hours, or in minutes, you have responses from people you've never met but who are interested in what you have said. You sent out a message in a bottle, and you got an answer. Your voice was heard!

At your computer, you are able to talk with people from around the world who share your most obscure or cherished or vitally important interests. You can join support groups (I found one called *alt.support.headaches.migraine* that has turned out to be a lifesaver). You can get gossip from highly placed sources, including famous people, who sometimes contribute to discussions (Douglas Adams, author of *The Hitchhiker's Guide to the Galaxy*, has been known to post to *alt.fan.douglas-adams*). You can buy rare books and videos (I completed my collection of rare paperbacks by the brilliant, manic, prolific "noted futurist" Harlan Ellison this way). You can fall in love (two of my best friends married guys they met on the net). You can be sure that you are not alone.

But venturing out into this brave new world as a woman on your own holds the same perils as it does in the nonvirtual world. Contributors to newsgroup discussions need not provide any information about themselves, which, you might think, would eliminate the gender imbalances that women run up against in face-to-face conversations. It is possible to send truly anonymous postings by using specially constructed mail-servers that strip off from a post all information about the author and place of origin. Thus, the possibility of physical intimidation (and danger) can be greatly reduced, and perhaps even removed. But those who are used to having control will try to keep it in any way they can, and in cyberspace that translates to *verbal* intimidation: insults, harassment, gratuitous nastiness and condescension. For a woman, it's like walking down a city street in a short skirt. As one woman told me about her experiences net-surfing:

> There are many times when people treat me completely the same when they find out I'm female, but many times I immediately get assaulted with pick-up lines or more subtle flirtation, condescension, or in some cases I am

completely ignored. And here's something interesting, several times woman have paged me completely out of the blue trying to pick up on me when they think I am male.

Let me back up a bit and explain how the newsgroups work and why they are so powerful.[1] Newsgroups like the ones I've mentioned are the Internet equivalent of BBSs, or bulletin board systems.[2] BBSs and newsgroups are indeed like physical bulletin boards: there are continuing topics of discussion, as well as random comments, and users can contribute or post as much or as little as they like. Even the word "post" conjures up the image of thumbtacking a message to a bulletin board. Some people do nothing but read the postings; this behavior is known as "lurking."

Since Internet newsgroups are available in almost every part of the United States (and many other places in the world), you have to imagine a bulletin board of infinite size, with many, many categories and subcategories. The nature of newsgroups is such that you cannot have real-time conversations with any of the other posters (to carry the analogy through, think of mailing in your message to the bulletin board, where a machine posts it in the order in which it was received), but when posts follow each other in order, it can seem to the reader much like a face-to-face exchange.

Topics of discussion are arranged as individual newsgroups. Within each group dozens of discussions may be occurring simultaneously (a continuing discussion on one particular topic is called a "thread"), but they are all supposed to be related to the general topic of the newsgroup.[3] For example, one of the newsgroups I read is *alt.comedy. british.* As groups go, it's fairly low traffic, with about forty new posts every day. A typical list of threads might look like this (I've deleted the authors' names in the interest of privacy):

1 Mr. Bean videos = which episodes?
1 The British-to-American TV Comedy Concept...List v0.91
3 General Email address for BBC
2 BBC fan mail
4 Chef! How many episodes?
1 Rise and Rise of Michael Rimmer

1 French and Saunders/AbFab skit question
1 Tom of "Waiting for God"

Subject lines are usually pretty descriptive, allowing you to choose the topics you want to read about and to ignore the others. The number preceding the topic tells you how many articles there are on that subject, so you know what you're getting into.

It's easy to search for newsgroups that might be of interest to you. If you want to find a support group for eating disorders, for example, you could do a general search for the pattern "support" and get results like this:

alt.support.anxiety-panic
alt.support.big-folks
alt.support.breastfeeding
alt.support.cancer
alt.support.depression
alt.support.depression.seasonal
alt.support.diet
alt.support.divorce
alt.support.eating-disord
alt.support.loneliness
alt.support.menopause
alt.support.obesity
alt.support.short
alt.support.shyness
alt.support.single-parents
alt.support.sleep-disorder
alt.support.stuttering
alt.support.tourette
soc.support.fat-acceptance
soc.support.transgendered

Often a search will turn up something you hadn't expected: For example, in the preceding list there is a newsgroup specifically for those with eating disorders, but *alt.support.diet* might also have useful information. I happened across the *.migraine* group by accident—it was one of the new groups being added that day, and it jumped out at me as it scrolled by with three million other new groups. It came just at the right time, too, as I was beginning to think I would never find any help. Once I joined this group, I found out why chocolate gives me migraines (theobromines), what the newest medication for migraines is (sumatriptan), and that other people are a lot worse off than I am (at least I never had to go to the hospital!). I felt connected. These were people who understood what I was talking about and who offered me not only sympathy, but good advice.

On the flip side, it's pretty discouraging to see that the highest traffic is generally on the *alt.sex* groups—and there are lots, with new ones being added daily (today I saw *alt.sex.torture*). At last count, the *alt.sex* hierarchy on my system showed about fifty-five different groups, ranging from *alt.sex.anal* to *alt.sex.wrestling*. It's a small percentage of the total groups available (around seven thousand), but there are more *alt.sex* groups than *alt.books* (seventeen) or *alt.politics* (forty-seven) groups (although fewer than *alt.tv* with ninety-one, and *alt.sports*, which has an unwieldy 121 groups. Expansion teams, I guess).

These *alt.sex* groups are on the whole by men and for men—guys looking for women to hit on, and guys posting endless fantasies (some of them are pretty offensive by anyone's standards). I've read and written a lot of erotic literature, and at first I thought some of these groups, like *alt.sex.stories*, would be a good forum for unpublished material; but rather quickly I found the *alt.sex* area to be boring (the writing is almost universally terrible) and kind of scary, too. I was afraid to post anything, even a response to somebody else's work, because I didn't want to be contacted by some of the creeps who posted poorly written sex-with-terrified-women stories.

After you join a group, it's best to lurk for a while and get the feel of it. Each newsgroup has its own culture and its own social conventions, and unless it is a brand-new group, you must be prepared to behave like the native population. This is not to say that all types of behavior are acceptable. I've met some of the most gratuitously rude people in my life on the net, people who slagged off anyone who didn't agree with their opinions about the world.[4] These people are usually men, and it's exactly this behavior that is a problem for women. Newsgroups are public places, and as you have probably noticed, men and women behave rather differently in public, especially when they feel strongly about the subject at hand.

Computer-mediated communication (CMC for short—

yet another abbreviation) is so new that very little quantifiable research on it has been done. Sure, you can pick up almost any newspaper and read anecdotal accounts of online experience, but we want numbers, hard data. Susan Herring, a linguist at the University of Texas, Austin, is one of the pioneers in data-driven research on gender and CMC, and her findings consistently support the notion that online communication is male-dominated and male-oriented. Generally, men tend to use strong assertions, self-promotion, authoritative orientation, challenge and sarcasm, while women use apologies, questions, personal orientation and explicit justification in their discourse[5]—in other words, men dominate the discussion and expect everyone else to participate on their terms.

Let me make clear that not all men and women behave this way. The problem with generalizations, especially those about gender-related behavior, is that there are always counter-examples. Yes, it's true that the behavior of men and women is almost identical when compared to the behavior of humans and cats; it's also true that we live in a patriarchal society and, therefore, women's behavior is still seen as marked, different from the norm. So the tendency of men online is to view their own behavior and opinions as the standard and consequently, to brand others as wrong.

As Herring puts it, "the metamessage communicated is I TELL YOU THAT YOU ARE WRONG."[6] This is verbalized in many different ways: through put-downs ("This discussion is stupid," "your pitiful example..."), name-calling ("idiot," "my dear puppet"), sarcasm ("Your insight is 'truly dizzying'"), personal insults ("Make up your mind—it shouldn't take a lot of work"), and authoritative claims ("I know for a fact..." "The truth is...").

In contrast, the metamessage of online women can be characterized as: I SUPPORT AND VALUE YOU, and I THINK THIS, BUT YOU MAY THINK OTHERWISE.[7] These messages are expressed as introductions and codas to the main part of the post: "Sorry to change the topic..."; "Thanks for listening to me rant..."; "I think what you said is very important, and I agree. But..."; "Sorry to bother you, and thanks in advance." These female-oriented behaviors are much more overtly polite than those generally used by males.

Herring also observes that "discussion on each of the lists

investigated tends to be dominated by a small minority of participants who abuse features of 'men's language' to focus attention on themselves, often at the expense of others."[8] This adversarial discourse—one that encourages conflict and argument—has the effect of intimidating women who want to avoid this kind of interaction. In subsequent studies, Herring has found that even though men *claim* to prefer less aggressive CMC, in actual practice they continue to act like bullies.

An example is the postings in the group *alt.feminism*, which had originally been started to discuss views about feminism, as someone mentioned in their short history of the group:

The idea of *alt.feminism* began in *soc.men*. There was a thread about the moderators of *soc.feminism* rejecting articles and the anti-man nature of many of the posts...Others wanted to create a group to reflect a more open discussion of feminism. Not just a forum for dogma...T.J.W. wrote a charter which simply stated that the group would be for all who wished to discuss feminism, both anti and pro. [post #8473][9]

By early 1993, the group had turned into a shouting match among ten men, all of whom felt that they were right and the others were wrong, wrong, wrong. Rarely did women post; when they did, they were the target of incredibly nasty follow-ups, as below in a discussion about gays in the military. Here is what one woman posted:

What disharmony will they cause? The people that are worried that a member of the same sex will be looking at them in the shower or coming on to them in the barracks should stop flattering themselves and start thinking about what their jobs really entail.[10]

One man replied:

Really "entail"! Quite a punny lady aren't you. I would consider true social justice to be when you get assaulted by a bulldyke named Bertha. Twice as big as you, she laughs as you suddenly realize how aggressive female homosexuals can be if they think they have an easy lay like a white liberal. Especially when the liberal no longer recognizes right from wrong or the implications of having to live in the amoral world she has tacitly created.[11]

Charming, no? And this isn't even the worst.

Let's look at how it's supposed to work. Here's a sample post to a fictitious newsgroup:

Newsgroups: alt.women.health
Subject: tampons
Summary:
Followup-To:
Distribution: us
Organization: University of Pleaseyourself, California
Keywords:
Cc:

Greetings all,

Does anyone know where I can get unbleached tampons?
I've heard that they pose less of a risk for Toxic Shock Syndrome
than the usual ones.

I'll post a summary of responses so reply to me by email.

Thanks!

Marlene Taylor
mtaylor@somenet.com

Your posting is then added to the group for all to see. Often your newsreader will be set up to automatically include the "article" you are replying to, so if I followed up to the above article, it would look like this:

Article 5864 of alt.women.health:
From: sutton@garnet.berkeley.edu
Newsgroups: alt.women.health
Subject: Re: tampons
Date: 26 Jan 1995 20:45:25 GMT
Organization: University of California, Berkeley
Lines: 14

On 24 Jan 1995 mtaylor@somenet.com wrote:

>Does anyone know where I can get unbleached tampons?
>I've heard that they pose less of a risk for Toxic
>Shock Syndrome than the usual ones.

I think you can get them in your local health food
store or purveyor of earth-friendly products. And if
they don't have them, you should ask them to order
some!

They work just as well as "regular" bleached tampons, BTW

laurel

You can post to a group as many times as you like, you can make your posts as long as you like, and you can say just about anything you want without being censored—overtly, that is. It may happen that no one responds to you, or someone may chastise you for posting:

Article 5865 of alt.women.health:
From: know-it-all@divine.cpm
Newsgroups: alt.women.health
Subject: Re: tampons
Date: 26 Jan 1995 20:45:25 GMT
Organization: Omniscient Com Inc.
Lines: 14

On 24 Jan 1995 mtaylor@somenet.com wrote:

>Does anyone know where I can get unbleached tampons?
>I've heard that they pose less of a risk for Toxic
>Shock Syndrome than the usual ones.

God! Don't you people read the FAQ? This topic has been discussed TO DEATH already. I'm so sick of clueless newbies who CAN'T EVEN BOTHER TO READ THE DAMN FAQ!

Thanks *so* much for wasting time and bandwidth.

One Who Knows

Okay, so this is an exaggeration, but not by much! I've seen this kind of reply almost verbatim in response to innocent questions and attempts at raising a new topic. How would you respond to this? Especially when it's possible that *thousands* of people might be reading that post. The accessibility of newsgroups is both good and bad. Because there is no one person in charge of a newsgroup, anyone can post—including those who might not otherwise be heard or be inclined to speak in a face-to-face (f2f) encounter. But it also means that people who have a lot of time on their hands can post and post and post until everyone groans at the sight of their names. Just as there is infinite possibility for support and encouragement, there is infinite possibility for abuse and humiliation.

Not every group is like this. For every net.group, there is

a different net.culture and different standards of behavior. Most of the groups I've chosen to read have a high politeness level and a lot of group solidarity: People support one another and are quick to initiate a "group shun" of loud-mouths or folks just trolling for trouble. I don't think it's a coincidence that these groups have a large percentage of female participants, and that the all-female mailing lists I belong to never erupt into fights.

In an unmoderated group, everything that is posted to the group automatically appears—a request might get twenty responses, or two hundred, or two thousand. If you were the original poster, you'd want to read those posts, wouldn't you? What if some of them were friendly and responsive, but the majority of the posters were just spoiling for a fight? How long would you bother to keep reading? Aren't there some rules in this place?

Well, yes and no. Think about it: No one ever tells you what the rules of f2f conversation are, except childhood admonitions to stop talking with your mouth full. We just learn it by observing, and it's very clear when we've made a conversational mistake—strained silences, icy glares, immediate change of topic. The fear and possibility of committing a social faux pas are strongest in an unfamiliar environment, especially when the conversational partners are strangers, strangers who have knowledge or power.

I said that conversational rules are rarely made explicit these days. This was not always the case: Until the mid-1950s in this country, etiquette books were regarded with respect and consulted often. A book such as Emily Post's *Etiquette: The Blue Book for Social Usage* (first published in 1922 and by 1955 in its eighty-second printing) provided detailed instructions on proper behavior for any occasion. "Courtesy manuals" were originally written (for and by men) in the Middle Ages for the new upper classes, providing them instruction on how to move into the non-communal culture of money and big houses without offending anyone important, like royalty.

Similarly, books like Post's served as handbooks for the nouveau riche of the United States, people who were unprepared to deal with all the intricacies of high society and needed to know how to behave without offending anyone important, like bankers and politicians. We could consider "netiquette" the newest version of behavioral guidance—prescriptive rules to keep newbies from offending established users. Books on netiquette abound; your Internet provider probably has a whole list you're supposed to read before signing on. A book like Virginia Shea's *Netiquette*[12] is typical of the rest—slim, attractively packaged and expensive, like all self-help books.

There may have been a time when net.culture was more homogenous, and cultural values were shared by the few who regularly used computers to communicate with each other. Even though there has always been a division between those who used the net for research and development ("scientists") and those who used it for their own ends or just for fun ("hackers"), the net has been, and continues to be in many ways, a masculine place, its atmosphere and protocol often being compared to the early days of the American West and the concept of "frontier justice."[13] But I don't think this metaphor works anymore, not with the explosion of online services and interfaces such as America Online, Prodigy, Delphi and so on. Rather than as the American frontier, we might think of cyberspace as a sprawling urban metropolis with people everywhere. Occasional conflict seems inevitable. And conflict often arises through mismatches of expectations and communication styles.

In f2f interaction, we use politeness strategies to avoid conflict and to punish offenders. But most of these reside in tone of voice, body posture and facial expression. Obviously you can't use those strategies in CMC, although emoticons (the little sideways smiley faces :) people sometimes use) are an attempt to mimic human features. Superlinguistic communication is replaced by linguistic dexterity and in some cases, linguistic overkill in the form of flaming—a sharp, brutal and usually insulting response to a post. Flaming is considered a type of "frontier justice," to return to the Western metaphor—a quick and dirty way of dealing with an alleged offender. Of course, the problem is that "offenses" can be anything from racist remarks to asking for a FAQ,[14] and anyone can find a reason to be offended if they look hard enough.

Those who would maintain order in cyberspace—system operators (sysops), conference managers, owners of discussion lists—prescribe extreme politeness to avoid conflicts, or at least to avoid miscommunication, with one exception—

flaming. This politeness is translated into user guidelines that instruct you to be brief, accurate, relevant and clear. That seems entirely appropriate, given the medium, and further, most netiquette guides are just updates of old etiquette guides. We're supposed to behave online just as we're supposed to behave in a cocktail party full of strangers. Yet flaming is still acknowledged as a way to deal with stupid or annoying behavior. Would you punch someone who accidentally stepped on your toe at a crowded party?

Why is flaming acceptable at all? The answer is that in cyberspace—as in society in general—the dominant group makes the decisions about what is appropriate behavior. The dominant group on the net (both in sheer numbers and in level of use), as in society in general, is male, and males see adversarial behavior as friendly; flaming is okay. Individualism is paramount, and the truth will eventually be shaken out of encounters built on direct conflict.

Almost all netiquette guides explicitly condone "appropriate" flaming; Shea has an entire chapter on "The Art of Flaming."

> Although flames often get out of hand, they have a purpose in the ecology of cyberspace. Many flames are aimed at teaching someone something (usually in overstated language) or stopping them from doing something (like offending other people). Flame messages often use more brute force than is strictly necessary, but that's half the fun.[15]

And even Emily Post does not criticize the androcentric style, claiming that arguments and fighting are okay as long as they are not personal: "Argument between cool-headed, skillful opponents is a delightful amusing game, but very, very dangerous for those who may become hot-headed and ill-tempered,"[16]—so net.style is just an extension of prescribed conversational style! The more contemporary Miss Manners, however, bemoans the fact that "arguing is now considered to be exchange of thought...."[17] The problem with flaming is that once any kind of flaming is accepted, it opens the way for every kind of mean-spirited insult in the guise of "an amusing game."

As in conversation, once combative behavior becomes

acceptable, some people cannot find other ways of expressing themselves: The one in greatest danger of making enemies is the man or woman of brilliant wit. If sharp, wit tends to produce a feeling of mistrust even while it stimulates. Furthermore, the applause which follows every witty sally becomes in time breath to the nostrils, and perfectly well-intentioned people, who mean to say nothing unkind, in the flash of a second "see a point," and in the next second score it with no more power to resist than a drug addict has to refuse a dose put into his hand.[18]

As Herring and others have shown, it is clear that women in general have a different style of CMC than men, one that is much more supportive and much less adversarial. So while men see flaming as friendly, cooperative and within the bounds of conversation, to women it is obnoxious—totally uncooperative behavior.

With the exception of condoning flaming, netiquette rules in general are basically rewordings of classic etiquette rules and seem to head off conflict. Here are a few examples:

> ***Etiquette*** *Ideal conversation is a matter of equal give and take. Try to do and say only that which will be agreeable to others.*[19]
> ***Etiquette*** *Conversation, which is supposed to be a two-way street, is treated by many people as if it were a divided highway. They may acknowledge that traffic must go in both directions, but speed independently on their own way, expecting you to do the same on your side.*[20]

> **Netiquette** Post and reply appropriately.[21] Don't send abusive, harrarassing, or bigoted messages.[22]
> **Netiquette** Treat others on the list as you would want to be treated.[23]

> ***Etiquette***...*a bore might be more accurately described as one who insists on telling you at length something you don't want to hear about at all.*[24]
> ***Etiquette*** *Conversation being an exchange, long stories, such as jokes or travelogues, cannot be included unless they are abbreviated and offered in illustration of the conversation's idea.*[25]

Netiquette Don't waste expert readers' time by posting basic information.[26]

Netiquette Don't make a posting that says nothing but 'Me, too.'[27]

Etiquette Those with vivid imaginations are often unreliable in their statements.[28]

Netiquette Make yourself look good online: Know what you're talking about and make sense. [29]

Netiquette Egregious violations of Netiquette: electronic hoaxes, rumors...[30]

Injunctions against misinformation are important because it is so easy to spread rumors and hoaxes; it seems there is a new "Make Money Fast!" posting to each newsgroup once a week. Jan Harold Brunvand, an expert on urban legends, has recently reported on the very new phenomenon of "faxlore," urban folklore spread through faxes and newsgroups.[31] Fake warnings about deadly computer viruses, too, are spread across the country in a matter of days, causing panic and confusion. Someone's paranoid delusions, which you would ignore if they came from the crazy guy at the bus stop, are raised forty points on the believability scale simply by virtue of being written down and transmitted by computer. Newsgroups, especially their FAQs, are seen as sources of reliable information, and vigilance against erroneous statements should be maintained.

Etiquette He who talks too easily is likely to talk too much![32]

Etiquette Try not to repeat yourself...[33]

Etiquette The chatterer reveals every corner of his shallow mind...[34]

Netiquette [when posting] don't automatically include the article to which you are responding...if you want to respond point by point, edit the discussion down so only the relevant sentences are included. [35]

Netiquette Format your postings nicely. Use a subject which is descriptive.[36]

Netiquette Keep paragraphs and messages short and to the point. [37]

Netiquette Acronyms can be used to abbreviate when

possible; however, messages that are filled with acronyms can be confusing and annoying to the reader. [38]

Netiquette Respect other people's time and bandwidth. [39]

Netiquette Don't rely on the ability of your readers to tell the difference between serious statements and satire or sarcasm. [40]

The prescription here is against wasting time and bandwidth through verbiage or ambiguity. The brevity guideline is continually violated in CMC because it is so easy to do—with a few keystrokes one can send many pages to people who have neither the time nor the interest to read them. In handling follow-up posts, some newsreaders employ automatic inclusion of all of the previous post, which may include many lines of citations of posts previous to that, and so on. Thus, though netiquette recommendations vary somewhat from guide to guide, they all recognize the importance of making CMC clear, descriptive and to the point.

You would think that with all these rules and advice there wouldn't *be* any problems online. But the laws are a reaction to the behavior, and merely saying the behavior is bad doesn't mean it will go away. In Real Life, smart women don't go down dark streets alone, don't flash expensive jewelry in public and always know where they are going. In cyberspace, the same principle applies: don't go alone into a dangerous area (like maybe the *alt.sex* groups!) and wear your flame-retardant suit.

Women have traditionally been physically excluded from the places where the important things get done—schools, governments, courts. It is only relatively recently that they have made their way into positions of power, often with great difficulty and under the threat of physical danger and harassment (Tailhook, anyone?). But Internet communities spring up and grow at a phenomenal rate, and you don't need a license to start one. *Women cannot be excluded from cyberspace!*

My purpose in writing this essay is to encourage women to venture forth on the net. Yes, rude behavior and sexism are out there, and they're annoying, but we don't have to put up with it. There are ways of dealing with someone's obnoxious behavior without being intimidated by it; most importantly, you have to avoid playing their game.

If you see a pattern of flaming or argumentative posts, you might respond privately and politely and ask the posters why they're being so aggressive; refuse to respond publicly; initiate a group discussion of standards of behavior; leave a newsgroup and start your own; and in the most extreme cases of harassment, get in touch with the system administrator and complain.

Some women have said that to survive on the Internet, women need to get used to flaming and to learn how to flame back: To take it and to dish it out, so to speak. This is just the old line about having to act like a man in a man's world. But why should we? I certainly don't want to start acting like an asshole every time I post to Usenet. I can acknowledge that there are flames in *alt.society.generation-x* without responding to them or posting a flame of my own; I can continue to carry on conversations with like-minded folks and have fun. I'm not going to let some stupid jerk spoil my cocktail party, especially when I can make him go away with one push of a button: The all-powerful Delete key.

The future is going to be online, and there has never been a better chance for women. Go girl! Get out to that cocktail party, and make sure they know you have arrived!

Endnotes

1. Of course, if you really want to know about how newsgroups work, you should go out and consult a serious book on the subject, like Ed Krol's *The Whole Internet User's Guide and Catalog,* second edition (Sebastopol, Calif.: O'Reilly and Associates, 1994).

2. Most BBSs are accessed by using a modem to dial up another computer. There is a limit to how many users can be connected to a BBS at any one time, and since the phone calls are charged to a phone bill like any other call, participation tends to remain local.

3. A lot of the fun in newsgroup discussions is in following threads as they stray further and further from the official newsgroup topic—in Usenet slang, WOTP (way off topic post), which requires an ObRef (obligatory reference). Warning: abbreviations and acronyms are a *big* part of newsgroups, IMHO (in my humble opinion).

4. Actual subject heading on the BH 90210 group: "YOU PEOPLE WHO WATCH THIS DUMB SHOW ARE A BUNCH OF LOSERS."

Some people have nothing better to do, I guess.

5. Susan Herring, "Gender and Democracy in Computer-Mediated Communication," *Electronic Journal of Communication* 3.2 (1993). See also Herring's "Posting in a Different Voice: Gender and Ethics in Computer-Mediated Communication," *Philosophical Approaches to Computer-Mediated Discourse*, ed. Charles Ess (Albany: SUNY Press, forthcoming).

6. Susan Herring, "Politeness in Computer Culture: Why Women Thank and Men Flame," ed. Mary Bucholtz, Anita Liang and Laurel Sutton in *Cultural Performances: Proceedings of the Third Berkeley Women and Language Conference*, (Berkeley, Calif.: Women and Language Group, 1994).

7. Ibid., 283.

8. Herring, "Gender and Democracy," 9.

9. Laurel A. Sutton, "Using USENET: Gender, Power, and Silencing in Electronic Discourse." *Proceedings of the Twentieth Annual Meeting of the Berkeley Linguistics Society* (Berkeley, Calif.: Berkeley Linguistics Society, 1994), 4ff.

10. Ibid., 4ff.

11. Ibid., 516.

12. Virginia Shea, *Netiquette* (San Francisco: Albion Books, 1994).

13. For more on this analogy and the history of cyberspace, see John Perry Barlow, "Crime and Puzzlement" (1990; available through FTP at *ftp.eff.org*, in the */pub* directory) and Bruce Sterling, *The Hacker Crackdown* (New York: Bantam Books, 1992).

14. Frequently Asked Questions. This is a list of the most common questions and answers relevant to a particular newsgroup. Usually one regular contributor to the group is appointed (by general consensus) Keeper of the FAQ and posts it semi-regularly. I warned you about these acronyms, BTW (by the way).

15. *Netiquette*, 71.

16. Emily Post, *Etiquette* (New York: Funk and Wagnalls, 1955), 48.

17. Judith Martin, *Miss Manners' Guide to Excruciatingly Correct Behavior* (New York: Warner Books, 1979), 167.

18. Post, *Etiquette*, 50.

19. Ibid., 45.

20. Martin, *Miss Manner's Guide*, 181.

21. Krol, *The Whole Internet User's Guide*, 150.

22. Ibid., 94.

23. Arlene H. Rinaldi, *The Net: User Guidelines and Netiquette* (Florida Atlantic University: Academic/Institutional Support Services Publication, 1994; available through anonymous FTP at *ftp.lib.berkeley. edu*, path: */pub/net.training/FAU/Netiquette.txt*.

24. Post, *Etiquette*, 49.

25. Martin, *Miss Manner's Guide*, 166.

26. Shea, *Netiquette*, 32.

27. Patrick D. Crispen, "Roadmap" (1994), § 7; available through email from LISTSERV@UA1.UA.EDU.

28. Post, *Etiquette*, 46.

29. Shea, *Netiquette*, 33.

30. Ibid., 85–86

31. Jan Harold Brunvand, "'Lights Out!' A Faxlore Phenomenon," *Skeptical Inquirer* 19:2 (1995), 32-37. This particular rumor warned that gang members would be driving at night with their lights off; if a passing motorist flashed their lights, the gang would follow the "good Samaritan" home and kill them. Brunvand traced the migration of this rumor around the country, as it was transmitted through faxes, email and newsgroup postings. The source of the rumor was never pinned down, and there were no reports of this type of attack anywhere in the country.

32. Post, *Etiquette*, 46.

33. Ibid., 46.

34. Ibid., 51

35. Krol, *The Whole Internet Guide*, 150.

36. Ibid. 150.

37. Rinaldi, *The Net*, 4.

38. Ibid., 5.

39. Shea, *Netiquette*, 32.

40. Crispen, "Roadmap," § 7.

41. Ibid., § 7.

"So Please Stop, Thank You": Girls Online

Michele Evard

Many people have noticed the gender split on Usenet newsgroups, observing that the vast majority of people posting messages are men. It has been informally estimated that less than ten percent of the public messages are written by women. This is much smaller than one would expect, given that an estimated 36 percent of Internet-accessing accounts belong to women.[1] Many women have allowed their voices to be drowned out, and both men and women have left the net entirely when they've become disgusted with flaming and other obnoxious behavior. It seems that on the net you have to either "put up or shut up"—is there truly no other option?

There might be. When working with children on a local network, I found that the girls did not avoid writing public messages. In fact, although there were only a few more girls in my project than there were boys, the girls wrote 58 percent of the messages. The girls who received negative responses did not back off, but held their own. Why were these girls so different from their adult counterparts?

Questions of representation became more important to me when I began designing an online environment for children to use in school. I wanted to design a space that would allow equal access to all the children involved. What could I do to provide a space that any child could use comfortably? I decided to use a computer-based newsgroup environment similar to Usenet, without any private email, and observe the children's behavior.

In the Classroom

In the spring of 1994, I introduced NewsMaker, a software system written by Mark Kortekaas which I helped design,[2] to three classrooms of fifth-grade students and two of fourth-graders. Because I believe that children learn by sharing their ideas about topics that interest them, I set up a few groups, but told the children they could start new topics at will.

The students could use NewsMaker to read, write and modify articles, but had no assigned newsgroup projects. Students in one advanced-work fifth-grade class were told they could use the network to ask questions about their current computer-based assignment, which was to design and create educational video games to teach younger students about the ocean.[3] The other advanced-work fifth-grade class could be online consultants for the new designers, since they had already done similar projects.

Although the children posted many kinds of messages to their newsgroups, most didn't use the system to hold conversations. The children who did were those designing and programming educational video games about the ocean. As mentioned, they got help from a second class, called the consultants, who had done a similar project.

These children interacted online over a period of four months. Their experiences discussing their school projects show what can happen when children are given the ability to talk freely online about things that are important to them. The story of one girl in particular demonstrates different ways the children interacted and how their interactions changed over time.

Renee's Game

Renee was one of the game designers. A quiet girl, she enjoyed both working on the computers and giggling with her friends—when she was alone with them. She wanted to create a game in which the player would control a fish, guiding it to the pots of gold that would reveal facts about the ocean. Anytime the fish crossed one of the barriers she created, the player would be asked a question about the ocean. In addition, she wanted to have a whale that could chase the fish. It would be made out of two shapes (to make it larger than the fish) and controlled either by the computer or by another person, depending on programmatic constraints. Of course, Renee didn't have the entire design set out on the first day; she, like most of the children, had plans that modified over time.

First Questions

On the first day of the game design project, a girl named Whitney had a question for everyone, so I showed her how to use NewsMaker to post it. When Renee had a question the next day, Whitney showed her how to use the system. Renee wrote:

March 15, 1994
shapes
How do you make two shapes move like one?

She did not sign this message, but since all messages had their author's name in the header, it was obvious who had written it. That day she received two replies from consultants (the other fifth-graders who had previously done a similar project). Suzanne's response read simply:

RENAE I DO NOT KNOW

Being used to Usenet conventions, I was surprised to read a message that said only "I don't know"; after all, the questions in a newsgroup are not generally directed to particular people, so what does it matter if one reader doesn't know the answer? This was only one of several such messages, many of which also included a possible alternative solution. In interviews I conducted prior to this project, several children had told me that they were uncomfortable when one of their friends asked them something they didn't know; it seemed as if they would "look stupid" if they admitted ignorance. In this online setting, however, it seemed to be a way for children to indicate that they wanted to help, but were unable to.

Ken, a rather loud, quick consultant, also answered Renee's question. This was another surprise, since it meant a boy was answering a girl's question. I had observed Ken's class during their computer time and noticed that boys would talk only to boys and girls would talk only to girls; I never saw a boy asking a girl a question or vice versa. Ken gave a matter-of-fact response that seemed to answer Renee's question.

Apparently it did not provide the information Renee wanted, however, because on the next day of the project, she posted this:

March 18, 1994
Following Shapes
How do you make one shape fallow another? One shape moving by the arrow or letter keys, the other moving by computer, following the other shape?
Renee

Albert's reply was yet another type of answer: a message that informed the questioner that someone else might have the answer:

Talk to Jorge. You two are in the same boat. (Pardon the pun)
Albert

Albert, a small and quiet boy, was the acknowledged computer expert in the classroom; he became very active online. Although he could probably have answered the question directly, as he did with many others, in this case he chose to redirect the questioner. This was probably more efficient; after all, Jorge had (with help) just solved a similar problem. Lisa also answered the question; since she was a good friend of Renee's, her sharp reply was somewhat surprising:

Renee,for the first thing you spelled follow wrong,for the second thing you do it by attaching the shapes.simple.ask me for details at lunch.
Lisa

This was the first "spelling flame" I noticed, although later in the project there were a few more. (It was not the typical type of spelling flame accepted on Usenet, as it was not typed perfectly, but it certainly was a start. No one responded with a flame for Lisa's typing.) I believe there were at least two reasons for Lisa's abrupt manner: First, she was generally rather tactless, and second, she and Renee had a comfortable friendship in which criticism could be taken as an attempt to help. Lisa's online communication was not particularly clear; her teacher said she often didn't take the time to write clearly in other settings either.

Other students gave Renee more details in posts they wrote during the next session. Unfortunately, they did not seem to address the question to Renee's satisfaction. Tina was a new student in the consultants' classroom; a tall girl who loved to dance and to laugh, she seemed to want to help even though she didn't have much experience. She wrote:

WELL YOU COULD USE YOUR FOUR TURTLES,THEN SEE WHAT HAPPENS. TINA H.-307

When I read this, and other messages like it, I was surprised; if these children had been asked, "How do I build a treehouse"? would they have said, "Use some boards and see what happens"? Renee did not seem to take offense, but replied by restating her question more clearly. Her reply demonstrated a type of reasoning I considered most important during this project: Instead of assuming that the respondent was stupid, or just "not getting it," some students would realize that what they themselves had written was unclear.

In real life it is very common for children of this age to assume that they are not at fault if someone doesn't understand their words. There is no need for them to clarify what they said; after all, they already explained it once, so the other guy must just be stupid. In their minds they have explained it; because they cannot review their verbal communication, all they have to go on is what they think they said.

In this online environment, however, children could go back and see what they had actually said. This happened many times during the project, and several children were surprised to see what they had really typed. I heard children say, "Oh, my question wasn't clear!" to others or just to themselves; another comment made was "I should've explained that I'd already tried that!"

Renee asked another question on the third day of the project:

How do you make words appear on the screen when you
reach a certain point of the screen or you reach a shape that is
on the screen?
Renee

Albert simply suggested she try a particular command. Several times during the project, Albert commented that giving someone a complete answer would not really help, since the person might just copy it and not learn anything. Other students didn't feel as strongly about this issue. Three consultants all answered in more explicit ways.

Renee's classmate Carrie, on the other hand, apparently took exception to Renee's question; she wrote:

RENEE I DON'T MEAN TO BE RUDE BUT YOU'RE ACTING DUMB
AND YOU'RE NOT THINKING!!!!!!!! GO TO YOU'RE TIME LINE
AND COPY YOU'RE PRECEDURE THAT DOES THAT!!!!!!!!!
UN NAMED

The capitalization was probably a Caps-Lock accident rather than an attempt at shouting, but the message was still quite pointed. It was also interesting to see that Carrie signed the message "UN NAMED" rather than just leaving her signature off. Her name would have been in the message header in either case, but writing "UN NAMED" indicated to me she wanted to comment anonymously, perhaps because she didn't want to be flamed back.

Carrie apparently later decided she shouldn't have criticized Renee so strongly, since at the beginning of the next session she repeatedly tried to delete the messages that included her comments.

Uncertainty in Renee's Answers to Others

While Renee asked several questions, she replied to quite a few as well. Many of her answers sounded as if they were written by a stereotypically uncertain young girl. During the session in which Renee posted her second and third questions, Albert wrote:

Does anyone know how to reverse a shape: make the black part

white, and the white part black?
Albert

Lisa replied simply "no," but Renee realized there was an ambiguity in the question. Instead of assuming she knew which question Albert meant, she responded by asking for clarification. After Albert replied, Renee suggested a solution:

Can't you just click space on all of the blocks in the square and then make the picture by clicking again? Then the black or background would be white and the picture or the white would be black?
Renee

This uncertainty was found more often in posts by girls than in those by the boys. It could be that Renee used it in this case because she was giving advice to the class's "computer expert," but she also seemed hesitant when answering a question by a girl who was not considered an authority:

Maybe you could color the box black as if you were writing words. I'm not sure if it will work, but it might,
Renee

Renee's style was even more apologetic in the following response to a question about musical tones:

I know this probably sounds strange, but do yu mean higher than tone h, because there is none. But if you want tone a,b,c,d,e,f, or g, then there is a proceedure in the box that has cards that has things you can do on the computer.   I'm sorry I don't know the exact answer right off hand, but go look. You might find it!

I do not want to suggest that all of Renee's statements were uncertain or apologetic, but several of them were. In any case, it was heartening to me to see that not only did Renee ask questions publicly, but she also spent time answering her classmates' questions in a constructive and supportive manner.

More Than Tech-Talk

Although most of the questions the children asked online concerned strictly technical issues, some children discussed the topics of their games and others talked about ways to make their games educational. Several weeks into her project, Lisa addressed a question about the style of her game to Renee. Renee's response was quite authoritative, even after the computer expert Albert had replied:[6]

April 29, 1994
RE: RE: FOR RENEE

> > RENEE, WHAT DO YOU DO ON YOUR GAME WHEN SOMEONE
> > GETS AN ANSWER WRONG?
> >
> How about printing on the screen, "Wrong!"
>
>   Albert
>
I agree with Albert, say "wrong." But you should also explain the correct answer, and sometimes put "wrong" in a funner way. For example, put nope, uh-uh, think again, etc.   Make it fun to be wrong!
Renee

Very few messages were directed to a certain person as this one was, and, as in this case, other students seemed to ignore the direction and answer anyway. Rather than contradicting Albert's suggestion or ignoring him as an annoying interloper, however, Renee responded in a way which incorporated his comment. It is also interesting to note that Lisa asked Renee a question online, rather than doing so in person, even though they are friends in the same classroom.

Another nontechnical topic that generated discussion was raised by a thoughtful girl named Tisha. She wrote:

April 29, 1994
what do you think?

Do you think that by asking trick questions you are teaching some one else?
tell me what you think
tisha

I was very glad to see the children asking for opinions on nontechnical aspects of their programs; most traditional school settings don't seem to welcome children's discussions of their personal pedagogical viewpoints. Renee answered Tisha during the next session:

I think that it depends on the trick question. Can you give an example? I think it might just confuse people with the correct answer, but give an example.
Renee

A week later, after a consultant named Kim answered, Renee responded again:

May 10, 1994
RE: RE: what do you think?

> > Do you think that by asking trick questions you are
> > teaching some one else?
> > tell me what you think
> > tisha
>
> I THINK THAT IF YOU HAVE TRICK QUESTIONS AND SOMEBODY
> IS TRYING TO ANSWER IT WOULDN'T BE RIGHT BECAUSE THEY
> ARE TRYING THEIR BEST TO ANSWER THE QUESTION AND THEN
> THEY'LL NEVER NOW WHAT IT WOULD BE. SO I THINK THAT
> TRICK QUESTIONS DO NOT TEACH SOMEBODY A LESSON.
> THAT'S MY OPINION.
>
> KIM E.
>

I think that you should have trick questions to add fun to your game. I think that it would be fun for somebody to find out that there is a trick question. It might make them think about the answers more carefully if you tell them before they start playing the game. But just be sure to tell them whether they chose the right answer, otherwise it might confuse them, and that wouldn't be good!
Renee

These girls along with the other three who responded were clearly focusing on the educational aspects of their games and paying attention to how the younger children would feel and what they would learn while playing the games. No boys wrote on this topic, perhaps because they were more focused on finishing their games than on helping other students by this point. Tisha and Renee had a conversation about trick questions in person after the above messages; Tisha told me she wanted to talk to Renee about it (even though they weren't friends) because Renee had asked her for examples.

Over time, the style of Renee's messages seemed to change. The ideas in her later messages were more clearly expressed than those in the early ones, and she gave more context for her questions. I had thought the children might stop using the online system when their questions became more complex, but that was not always the case. Renee's game progressed much as her messages did. She was able to work out her problems and produce a complete game that other children liked. Her game was the only two-player game in the class; she made a whale which a second player could use to chase the first player's fish. If the whale was able to catch the fish, the game ended, even if all the pots of gold had not been retrieved.

Participation by Gender: Is it Just Renee?

Given only the messages quoted above, it would seem that the girls in this project wrote much more than the boys did. Would the story be different if I had chosen another child? Certainly. Does this story present an overly biased picture? I don't believe it does.

Although Renee did write a high number of messages compared to the average for her class (she wrote twenty-three and the average was twelve), she was not the most prolific poster. For example, Albert wrote forty-one messages, Emilie authored thirty-one, Stephan twenty-three, Carrie twenty-two and Lisa twenty.

Of the 235 messages written by the twenty game designers, 138 of them were by girls. That means almost 59 percent of the messages were written by the ten female class members. There were ten girls and eight boys in the consultants' class; eighty-six messages were from girls, and sixty-six from boys. This class was more balanced (female consultants wrote an average of 8.6 messages each, and the boys 8.25), but the girls still wrote more. These few statistics show that girls participated in this online environment at least as much as the boys did, and in many cases, more than the boys.

Duh! Rudeness Online

Most of the time, neither boys nor girls were rude in their online messages. Although there were some striking exceptions, none of the messages approached the nastiness of common Usenet flames. The message that Carrie wrote and then tried to delete was the strongest message any of the girls wrote. There were also a few blunt ones, such as Emilie's "Renee, just go to your timeline!", but these weren't considered rude by the children.

Messages with "Duh!" in them, however, were considered rude by many children. For example, Ken sent this message in response to a question Carrie had posted:

March 15, 1994
RE: Shapes
> How do you make a round circle using two shapes?
>
  Make a semi-circle on one shape and another semi-circle on the
other.   DUH!!!

Don, another consultant, chimed in "Yeah!" Carrie replied by writing the only message from a girl that said "duh!" to someone:

Don I tried that and it made a skinny oval DUH!!!
CARRIE

I was glad to see that Carrie didn't just back off and consider her question stupid because of the responses she received in public from these boys. Some of the designers told each other these boys' messages were rude; I didn't hear anyone say they thought Carrie's response was also rude. Although four of the boys sent this kind of message very early in the project, the others were generally helpful and nonoffensive. One of the girls, a designer, felt strongly enough about rude messages that she wrote this:

March 21, 1994
Rude!
I think that some of the answers that are given are rude and
impolite. NewsMaker is not a place to talk about what happens
during the day, it is a place to ask questions and get answers.
Some of the answers that we are geting are rude and the people
that write them should stop. People should also stop answering
question impolitely. I think that if someone asks a question people
should not answer the question if they have nothing to say. They
should also not end the question writing something like "Duh!"

they should answer it with something like "And that is how you do it.". So please stop thank you.

Whitney

Three girls responded to this post, agreeing with Whitney. The only boy to reply was Ken, who only typed lines of random characters. He had posted various derogatory comments in other groups, as well as nonsense such as this, but none of his remarks were ever answered, and those to whom they were addressed continued to participate. Although Ken was certainly not supportive of Whitney's message, he didn't flame any of the girls involved, nor did he ever write "duh!" again. As a matter of fact, none of the children ever wrote "duh" or anything similar in their messages after Whitney's post.

Emilie dealt with the "duh" responses to her questions in a different way. This girl didn't take smart talk in person without some response, and she didn't online either. She posted this question initially:

DOES ANYBODY KNOW HOW TO MAKE DIFFERENT MUSIC THAN TONE C ,TONE B,...

EMILIE B.

Two of the male designers answered her the same day. Mark stated:

Nooooooooooooooo! Duh

And Stephan wrote:

Nope!! You can't. DDDDUUUUUUUHHHHHHH!!!!!!!

Emilie did not respond to their messages directly, but immediately put up a second question on the same topic:

(STEPHAN + MARK DON'T ANSWER THIS)
DOES ANYBODY KNOW HOW TO MAKE MUSIC ON THE COMPUTER????????????

There was no way for Emilie to automatically disregard messages by certain people (other systems have "kill files" and "bozo filters" that allow users to hide messages by particular authors, but no such mechanism was available to the children), but she found her own way to remain a participant without flaming back.

. . .

Breaking Some Stereotypes

There are many stereotypes about males and females that one might have expected in this setting: For example, boys are often rude, girls are more social, children only help their friends, girls get upset by obnoxious behavior and withdraw...the list goes on and on.

I was glad to see that many of these stereotypes were broken in this online environment; the children did not necessarily act in ways that simple stereotypes would have predicted. The boys were generally helpful, both to other boys and to girls; only occasionally did they write anything rude. The girls did pay attention to social aspects of being online, even to the point of writing about types of behavior that they considered appropriate, but they did not confine themselves to the social sphere; they also asked and answered technical questions in the public forum. Both boys and girls answered questions of children of both genders, whether or not they were friends. Many children were willing to admit when they didn't have an answer, and generally suggested alternative questions or possible answers. And girls did not simply withdraw when they received or read rude messages.

Other aspects of the children's interactions were also striking. For example, girls and boys answered about the same number of questions in both classrooms; however the girls were more likely to qualify their answers, while the boys were more direct. The girls more often used stereotypically feminine forms of writing, making suggestions and employing apologetic tones. Female designers asked twice as many questions as their male counterparts did, and while the female consultants asked a total of four questions, none of the male consultants asked a question. Perhaps some boys felt they would lose face by asking questions, but the girls seemed to view asking questions as a social activity rather than an admittance of ignorance.

My original questions asked if I would be able to provide a space that would not exclude any children because of their gender, ethnic group, clique or ability level. I also wanted to have a space where the children would be free to express themselves on important topics not typically supported in schools. I believe that for these children I was able to reach these goals.

The context of the children's communication was

critical to the success of the project: The fact that the game designers were each creating something that was personally important and that they would be able to share with others in the school meant that each one had a personal interest in seeing their communication succeed.[7]

The public nature of all messages may well have caused many children to think twice before writing anything offensive. A child who posted a flame might have been flamed in return—and not only by the child he or she flamed, but by other members of the community. Although teachers were not major participants in the project, it was clear that they could read any of the messages. In addition, the messages were not anonymous, and the students probably realized there could be negative consequences for unacceptable behavior. Since this network was local to the school, all of the children on it could locate anyone who had written a message. Because the author's real name appeared with every message, everyone else knew who each author was and could find him or her in person. This could be very positive, as in the case where Tisha and Renee talked in person after exchanging messages online, but it could also mean that if a message was offensive, other children could talk to the author in person about it.

Another aspect of the network being in a common physical community was that children may have taken others' real-life personalities into account when reading their messages. For example, Renee might have been offended by Lisa's spelling flame message if she hadn't known what Lisa was like in person. Children in this environment knew that there was a real person behind each message and that there were real repercussions of the messages.

Even given all of the above considerations, it still may seem surprising that boys and girls used the system in similar ways. Many people think of men and women as having vastly different interactional styles online. Some of the stereotypical behavior did exist in my project, but many of the children went far beyond what might be considered gender boundaries on the net. And the gender participation levels were clearly different from those on Usenet.

I believe that part of the reason for this happy occurrence was that all of the children started using the system at about the same time, and at the same age. If the boys had been online weeks before the girls were, it could have felt like a boys' activity and the girls may have acted differently— and vice versa. Of course, as we know, most of the early Usenet participants were male.

If the children had been a few years older, entering the crisis time of adolescence, their experiences may have been vastly different. Psychologists who have studied children as they become adolescents have found that girls seem to be more strongly affected by adolescence than boys are.[8] Perhaps the girls in my project wouldn't have participated as freely as they did if they had been dealing with the new set of social expectations and insecurities brought on by adolescence.

There are many uncertainties in any project like this, and the reasons for the children's particular interactions may never be entirely clear. The important point to me is that all the children were involved; neither girls nor boys let others chase them off. All of the children seemed to want to participate in some way.

Will It Be Different For Them?

Will these children's online experiences at school affect their experience when they jump out onto the net? What could, or will, they and others like them do to change Usenet?

If these children had started their online experiences with Usenet itself, rather than the local Usenet-style system in the school, I believe they would have acted much differently. For example, they would have seen flame wars and learned that rudeness is so common as to be acceptable. By giving them a local system, I provided a way for them to determine what their online community would be like, rather than requiring them to learn some other people's standards.

Now that they have been able to make that choice and see one type of community, what will they think about the net? The optimistic side of me likes to imagine that with an influx of students who have used local networks in ways I describe, the net will change: The percentage of obnoxious flames will go down as more and more of these children join; they will know they don't have to shut up to avoid being made targets; and they will tell those who are rude that their behavior is inappropriate. And the flamers will step back and think about what they are writing and cease their bullying ways.

However, after reading a nasty flame, or getting offensive email from some guy I have never heard of, I am more likely to think that newcomers who enjoy being offensive will take one look at the net and feel it is a place where they can be as rude as they want without any consequences. Those who don't like such behavior will quickly learn that they can't avoid it and quietly bow out.

After the experiences Renee and her schoolmates have had, I would hope that they wouldn't just give in to the flamers. Although messages such as Whitney's might have no effect on the Usenet community itself, because of its history, there are alternatives such as Emilie's—ignoring the flames and those who write them. These children know that those who post offensive messages don't have to do so to get their points across, and will not, one would hope, let rudeness affect their own actions.

The children I worked with have demonstrated that it is possible to have online discussion groups that are balanced across such lines as gender, race and academic ability—and that the online world is not inherently a hostile one.

Endnotes
. . . .

1. John Quarterman of Matrix Information and Directory Services, based on a survey in December 1994 which showed that 36 percent of Internet-accessing accounts were owned by women. Cited in Elizabeth Weise, "How Big Is It?" The Associated Press, 12 May 1995.

2. Mark L. Kortekaas, "News and Education: Creation of the Classroom Chronicle" (Master's thesis, MIT Media Arts and Sciences Program, 1994).

3. For more information about game design projects, see: Yasmin B. Kafai, *Minds in Play: Computer Game Design as a Context for Children's Learning* (Hillsdale, N.J.: Lawrence Erlbaum Associates, 1995). More information about this project in particular can be found in: Michele Evard, "A Community of Designers: Learning through Exchanging Questions and Answers," in *Constructionism in Practice: Designing, Thinking, and Learning In a Digital World*, eds. Y. Kafai and M. Resnick (Hillsdale, N.J.: Lawrence Erlbaum Associates, 1996).

4. Although all of the children's names have been changed, every

message I quote is otherwise exactly as the student typed it.

5.   As you can see, the children sometimes used all capitals; I believe this was because of their lack of typing skills rather than any desire to shout. Writing was difficult for many of them, and no one tried to force the children to make each message picture-perfect.

6.   "RE:" in the header of a message indicates that the message is a response or reply to a message with the same title. Thus the double-RE: here indicates that Renee is responding to Albert's response to the message entitled "FOR RENEE." The > > symbols preceed quoted lines.

7.   This theory is explained in more detail in Seymour Papert, *The Children's Machine* (New York: Basic Books, 1993).

8.   For example, Carol Gilligan, Nona P. Lyons and Trudy J. Hanmer, eds., *Making Connections* (Cambridge, Mass.: Harvard University Press, 1989), 15.

TEXTUAL REALITIES

MUDder? I Hardly Know 'Er!
Adventures of a Feminist MUDder

Lori Kendall

> You see a beautiful, pale-skinned young white girl.
> A flimsy negligee hangs half-open from her shoulders,
> revealing the first dim red half-circles of her aureolae.
> She's wearing slight white lace panties with a little pink
> bow at the rise of her mons veneris. Her dark red eyes
> glower at you like volcanic coals, and in her left hand is a
> four-foot stick bound in black iron bands, tipped with a
> steel ball studded with sharp, well-used spikes.
> —"Amnesia" character description from BlueSky MUD

How meaningful is gender online? On most current online forums, all communication is through text. Theoretically, people can name themselves and describe themselves however they choose. Does this ability to hide identifying characteristics "level the playing field," creating new opportunities for equality among male and female participants? Does gender identity become more fluid under these circumstances? Those were some of the questions I had a year ago when I started researching MUDs, which are online, text-based interactive forums.

I've been on BlueSky, a social MUD, for about a year.[1] I've made several friends and have come to like the rowdy atmosphere and scatological humor there. Sometimes, I sit laughing hysterically at my computer until the man I live with comes to read over my shoulder, wanting to know what's going on. This has led to jokes between us that I'm "turning into an eleven-year-old boy." And yet, if I hadn't had a re-search agenda—if I hadn't been motivated to figure out what was going on online no matter how offensive I found it—I never would have stayed on BlueSky or, for that matter, on any of the other MUDs I've visited. The rudeness and obnox-iousness I've been exposed to, the constant references to "babes," the penis jokes, the blow-job jokes, and so on, would have chased me off long ago, and they still sometimes weary me. Most women don't have my research agenda, and most are unlikely to find much of interest to them on MUDs, un-less they are science-fiction fans or, like me, have an unusu-ally graphic sense of humor or a high tolerance for people who do. Even these women are likely to find most MUDs unwelcoming.

But, as in my case, women who stick it out may find friends and other worthwhile contacts online. Perhaps more important, it is worth considering what these online forums will mean if they remain male-dominated. Given these con-siderations, I discuss here some of the ways in which gender operates online.

How MUDs Work

The acronym "MUD" originally stood for "multi-user dun-geon." Some MUD participants also now translate it as "multi-user dimension" or "domain." As the original termi-nology suggests, MUDs started as interactive role-playing games similar to Dungeons and Dragons. Currently, hun-dreds of MUDs are available on the Internet, and many still provide an environment for this type of game; people (in some cases, hundreds) can log on to the computer on which the MUD program is running and interact with each other and with the computer, solving puzzles and staging battles. Many MUDs, however, emphasize types of role-playing that are more like interactive theater. These MUDs are frequently based on popular science-fiction and fantasy works such as *Star Trek*, Anne McCaffrey's *Dragonriders of Pern* stories and Roger Zelazny's "Amber" series, among others. Still other MUDs are loosely organized around various imaginary themes, but are less games than settings for getting together with others and socializing. A few MUDs are also being de-veloped for use as professional and academic spaces. My comments are based primarily on observations of social MUDs, where the main activities are hanging out and chatting.

People interact on MUDs through "characters." Characters are programmed objects that act as surrogates for the participants. Each character has a name and description, as well as a gender "attribute," all of which are chosen by the participant. Descriptions are available to other participants using the *look* command. (A sample description is Amnesia's at the beginning of this essay.) The gender attribute allows the MUD program to assign the appropriate pronoun to the character in text referring to that character.

All interaction with the MUD program and with other participants on the MUD takes place through typed text commands. On the social MUDs, the commands used most often are those that enable the participants' characters to "speak" and "act"; the text of these communications is sent to the screens of other participants. The command to speak is known as the *say* command and the action command is called *pose* (or sometimes *emote*). Most MUDs are composed of several "rooms." Each room has a description and may contain several programmed "objects." Short commands, usually compass directions, enable movement from room to room. Here is a sample of MUD interaction. My character is called Wombat on this MUD; the commands I typed are after the > prompt symbol. (Note that when I use the *pose* command, the MUD program inserts my character name, Wombat, but when I use the *say* command, on my screen, the program inserts the pronoun "you.")

> look
Corridor
There are stairs going down to the west here and the corridor continues to the east. There is an unmarked door to the south. A door to the north leads to the guest bathroom, and to that door's right is a narrower door that probably goes to a linen closet.
> west
Stairwell
The corridor from the west ends here with short flights of stairs going up and down to the east. South leads to one of the master bedrooms.
Changer is here.

```
Changer says "hello.."
> pose waves
Wombat waves
Changer says "so,. whats everyone up too?"
> say not too much
You say "not too much"
```

Participants usually see only the text generated by people in the room they're in, but it is possible to *page* messages to people in other rooms on the MUD. In addition, the *whisper* command allows one to send a message to someone in the same room, but keep it hidden from other participants. Text may also be generated by interactions with programmed objects, using commands like *open box*. The effect of all these features and commands is to create a feeling of space—a virtual location—in which several participants can communicate with each other and interact with the program.

Objokes

MUDs can be very disorienting at first. During busy times, text scrolls by on your screen at an alarming rate. In the party atmosphere that usually prevails when many people are in the same MUD room, several different conversations may be going on simultaneously. Unlike at a physical party, conversational groups don't separate off into different corners of the room. All conversations occur in the same "space," and unless whispered, are audible to all in the room. Thus, different strands of talk intertwine in bizarre juxtapositions that can be extremely difficult to make sense of without practice.

In addition to the general disorientation, MUDs may feel particularly foreign to women. On some MUDs, portrayals of females as sexual objects become part of ritual-like interactions. These rituals are important because they demonstrate belonging to, and shared history with, the group. On BlueSky, for example, a common form of word play is the "objoke" (for obligatory joke). Objokes are puns based on word endings. Here is the standard (nonsexual) response to a participant's use of a word ending in -*us*:

```
McKenzie says "what's the rumpus?"
Parker says "rumpus? it nearly killed us!"
```

The more common form of objoke (probably in part because the word ending is more common) is based on words ending in *-er*. There are two standard responses (in the first example, the . o O (notation indicates a cartoonlike thought bubble):

Stem4 says "Pedants are sticklers for correct spelling."
elflock . o O (you brought 'er, you stickler)

Ronald Ann just gets her cable through the heater vent from the apartment downstairs... no choices
henri says "HEATER VENT"
Mender says "heat 'er vent? I hardly know 'er"

The latter response is so common that most participants do not complete it, leaving the remainder of the joke to be called up in the minds of the others. Here, two participants race to be the first objoke respondent:

Parker says "it's all the Badger's fault"
elflock says "Badger, i hardly"
henri says "badger why I hardly"
elflock winz
henri loses

The dominant motivations for objokes are probably the satisfaction of engaging in punning wordplay, and the use of an element of group subculture, in this case a running joke, to maintain group ties. However, the joke also evokes the image of a generic female sexual object (and not a particularly desirable one at that—note that the "you brought 'er, you _____ 'er" response implies the passing around of an unwanted toy). This can make BlueSky an uncomfortable, if not downright hostile, environment for women participants.

Being Female on a MUD

Far fewer women than men participate on MUDs, although the exact ratios vary from MUD to MUD. As in other areas of the Internet, participation by women is probably increasing, but newly arriving women encounter a social environment and behavioral norms formed largely by men. In some cases, these norms may be disturbing enough to discourage further participation by women. As one MUDder explains:

Women get treated differently from men. It's not that they get more slack or anything, but they get chased out differently. Part of it is that most women tend not to talk immediately, just by normal socialization. Guys are much more likely to mouth off early. But women often get turned off by less nasty hazing than men.

On the various Usenet newsgroups and email lists devoted to discussions of MUDs, MUD participants mainly worry about the effects of sexual harassment on women participants. The following stories are typical:

I recently (last week) showed a female friend how to MUD. ... We were sitting side-by side in a comp. lab during her 1st session.. and although she used a more or less male name. She marked her sex as female. The male character who soon found it his mission to pamper her in the game was very nice to her. All seemd well, and I went back to writing something in emacs on my screen.

After about 15 minutes (and 30.. maybe 40 newbie questions from my friend ;) This guy in the game started making advances on her..

A female friend of mine lagged out on one MUSH and another player took advantage of the situation and "posed" himself as raping her. When she unlagged, the person in question was gone and she was left with the equivalent of an obscene phone call on her screen.[2]

Generally my reaction was to foofoo that there was any problem except that I recently was able to get my wife interested in mudding. I could NOT believe what she had to put up with just to play. I got very angry at what seemed like a bunch of children trying out 'social' commands that they wouldnt ever do in public. It went far beyond role-playing.

Although it's unclear how prevalent this type of harassment is, it's only the extreme end of a range of behaviors directed at female characters (who, it should be kept in mind, may or may not be operated by female participants). Participants running female characters may find themselves the focus of constant sexual interest and innuendo.

MUD participants also describe other forms of differential treatment, especially overly helpful treatment of female characters:

Beryl whispers "the main area you notice it is when newbies log on. a female newbie will be given more breaks & more attention, even (especially?) by women"

In this situation, gender becomes almost a stance of relationship to the MUD—a way to designate one's desired treatment, rather than a statement of identity. Thus, some men portray female characters because they are dissatisfied with the "neutral" reactions they get when portraying male characters:

Amnesia whispers "'Oblivious' was my male persona briefly, but it was less fun"
You whisper "less fun? how so?" to Amnesia.
Amnesia whispers "hard to say. Perhaps less attention is paid male characters."
You whisper "hmmm. I've heard that from other people as well." to Amnesia.

Others specifically portray female characters for the sexual adventure of engaging in TinySex with other male characters.[3] This led one participant to comment:

I think the rule should be: If you are a homophobe, don't have tinysex cuz that cute broad might be a guy in real life. If you aren't bothered by this, have fun.

Meanwhile, many women portray male characters to avoid the very behaviors these men are courting.

Amnesia stares at CH. "Oh, no, you don't actually admit to being female on line, do you?"
Amnesia notes that she didn't mean to imply that CH was not a female, but rather that anyone who really is female canonically ought to pretend to be male

It is true that any woman who doesn't like the way she is treated when she logs on as a female character can change her identity to a male or neutral representation. However, this solution leaves intact a status quo in which being female, especially if you are new, means being hit on. Many MUDders, male and female, see this as a problem and are working to change it.

But differences in treatment of male and female characters are only part of the story. Even on MUDs where everybody is treated more or less the same, norms of behavior, as well as topics of discussion, contain gendered meanings and expectations. Here for instance, performances of swaggering masculinity by anonymous guests are accompanied by intimations of homosexuality used as put-downs:

Blue_Guest says, "I said, "Are there any EXTREMELY DESPERATE women in the house?"
Titwillow [to Blue_Guest]: and i asked why you wanted to know!
Blue_Guest says, "For my own info. dude!"
Green_Guest says, ".......Mike Lloyd is gay.....he sucks Steven Bruckner....."
Ebony_Guest says, "Fuck all of you muthafuckas"
Blue_Guest says, "Eat me Blissful, you must be a fag!"

In the milder example below, friendship between two male participants is expressed through a joking negotiation for the possession of female sexual objects. This conversation took place in a public room in which women participants were also present:

henri says "[Mender], if we meet a couple of supermodels in NYC, the rule is: you take the brunette, I take the blonde"
Mender says "what if they are both blonde?"
henri says "you take the shorter one"
Mender says "hsm"
henri says "you are shorter than me after all"
Mender says "OK"
fnord says "what if you meet a short blonde and a tall brunette?"
Mender says "trouble"
henri says "hair color overrides height"
henri says "domehead and I used to use the blonde/brunette system when scoping babes (from afar) at basketball games"

MUD Cultural Intersections

Puss-n-Boots is a beautiful, female felinoid. She stands about 5'3" and has beautiful cream colored fur all over her nicely curved body. She is a seal point Siamese, so her coloring is accented by dark brow thigh-high boots and dark brown elbow length gloves. Her Asian features are accented by her bright blue eyes. You notice in the top of her boot, an ornamental Tanto knife, from her native homeland.

You see, this is her newest form, and this time, she has no mortals to answer to. Around her neck is an Italian 500 lira coin given to her by someone very special to her. As you continue to

stare, she looks at you and whispers seductively, "You like what you see?"
— Puss-n-Boots character description from AniMUCK

MUDs form a subculture that intersects with several other subcultures, most of which are also male-dominated. These include the role-play gaming subculture; comic book fans (including particularly strong connections with fans of Japanese comics); science-fiction fandom in general; fans of particular elements of science fiction, such as the stories upon which many role-playing MUDs are based; and Furry fandom, around which several social MUDs are organized.

AniMUCK is an example of a MUD that caters to Furry fandom. AniMUCK generates considerable controversy in the MUDding community for several reasons: (1) many character descriptions are overtly sexual; (2) the Furry theme, derived from comics, science-fiction stories, and artwork involving anthropomorphic animals (known as Furries), is considered silly and perverse by some MUDders; (3) TinySex is rumored to occur frequently on AniMUCK; and (4) AniMUCK has been featured in several sensationalized popular press articles about MUDs, which many MUD participants feel paint an inaccurate picture of MUDs and MUD behavior in general. Flame wars concerning AniMUCK and Furry fans are fairly frequent on the Usenet newsgroups devoted to discussion of MUDs. Most of these flame wars are initiated by a few individuals, often in reply to newbie questions regarding AniMUCK. However, once initiated, the battle is enthusiastically joined by Furry supporters and detractors alike. Despite frequent pleas from newsgroup participants to let it drop, these battles can go on for weeks, only to be rekindled again a short time later.

Although AniMUCK is very different from BlueSky and ostensibly has a less masculine atmosphere, the status of women there is similarly problematic. Some of the controversy concerning AniMUCK simply reflects the fact that sexual behavior, rumored to occur there frequently, is controversial. However, the charge of sexism on AniMUCK also occurs. This charge frequently relates to the larger community of fans of Furry art and stories, not all of whom participate in MUDding. As one female participant commented in a posting to a Usenet newsgroup:

I was at a Worldcon [a science-fiction fan convention] a few years ago, and a Furry Party was widely advertised. Out of curiousity I walked in the room, and I'm afraid it was out of repulsion I left. ... Most of the artwork was of pornographic nature, women with huge breasts in sexual poses, sometimes with men (with horse-sized genitalia— i wouldn't mention it but it was so memorable!) involved in the pictures. I left the room, gasping from the cigar smoke but mostly from the intense anti-feminine atmosphere.

Although several Furry supporters commented on this posting, suggesting that the described gathering was atypical, my perusal of *non*pornographic Furry cartoon art available online by FTP (there are also examples of pornographic Furry art available) tends to support this analysis of Furry fandom. The male anthropomorphic characters depicted in the graphics files range from "dark-and-mysterious" wizard types to "happy-go-lucky-catboy-and-his-computer" types. None of these male characters are overtly sexualized. The female characters, on the other hand, have a much smaller range of variation and are almost always large-breasted and seductively posed. In contrast to the range of tools shown with male characters (some hold nothing, others are depicted with staffs, computers, etc.), almost every female character is wielding a gun, sword or similar weapon. This juxtaposition of large breasts and pointed weapons gives them a hypersexual "phallic female" appearance.

This is not to suggest that all participants on AniMUCK are sexist, or even that sexist behavior is encouraged. However, at the least, the cultural intersections with Furry fandom create a MUD cultural norm in which the female gender identity in particular is highly sexualized. Thus AniMUCK, like BlueSky, adopts a norm and tone of interaction largely set by men; one which is limiting, if not uncomfortable, for women.

Are You Male or Female?

Choosing a gender-neutral or male character may free a female participant from fears of direct harassment or overeager sexual interest, but regardless of the gender of her character, a female participant observing the types of conversations previously related is continually reminded of the male-dominated environment in which she moves. Furthermore, choosing one gender or another does nothing to change the expectations attached to particular gender identifications.

In theory, choosing a neutral gender designation would mean escaping the dualism of male and female gender expectations. (After all, in our face-to-face interactions, we don't meet many people we would designate as gender-neutral, and therefore we don't have a clearly assigned role for such persons.) In fact, it would appear that a significant number of MUDders use neutral designations just for this purpose. GammaMOO, for instance, has the following choices: neuter, male, female, either, Spivak, splat, plural, egotistical, royal, and 2nd. "Neuter" designates the character with the pronoun *it*. "Either" uses the *s/he* and *him/her* convention. "Spivak" uses a set of gender-neutral pronouns such as *e* and *em*. "Splat," similarly, uses **e* and *h**. As might be expected, "plural" uses *they* and *them*, "egotistical" uses *I* and *me*, "royal" uses *we* and *us* and "2nd" uses *you*.

As of October 17, 1994, out of 8,541 total characters on GammaMOO, 23 percent (1,991) were neuter, and only 2.5 percent (216) were anything else, other than male or female. Significantly, only 21 percent (1,770) of GammaMOO's characters were designated as female. Thus, male characters outnumber all others put together, and more people choose some form of neutral designation than choose female. But almost all the gender-neutral characters I have met were guest characters who had not yet set their gender designation, so it is unclear to me how well this strategy works. My guess is not very well, for the very reason that we *do* expect everyone to be either male or female. No one encountering someone using the pronoun *e* is likely to believe that this expresses their "true" gender, and is thus likely to treat the character's gender designation as a mere mask. Some may respect this desire to "hide" gender, but others probably will not.

Consider, for instance, that at least in areas where guests or other anonymous characters are common, the question "are you male or female?" is frequent enough to have acquired joke status among experienced MUD participants.

Copper_Guest says, "Infrared, Are you male or female?"
Infrared_Guest says, "male"
Copper_Guest syas, "Cool! Lets chat. So, you're from Austin, right?

Infrared_Guest says, "copper are male or female"

Copper_Guest says, "I am most definitely female."

Yellow_Guest [to Copper_Guest]: *most definitely*?

Copper_Guest says, "Just wanted to stress the fact, since you can't see me and all..."

Furthermore, since everyone knows that character gender need not reflect the face-to-face gender of the participant, setting one's gender doesn't make one immune to the "male or female" question.

Previous asks, "are you really female or is that just your char?"

achina [to Previous]: that question kind of surprises me. Why do you want to know?

Previous smiles at you.

Previous says, "just checking"

Previous says, "best to catch these things early.. people here tend to switch sexes almost as often as clothing"

achina is female in real life.

Previous says, "good :)"

achina still isn't sure why you need to know my RL gender, though.

Previous says, "I don't like being switched genders on, so I make sure early on so I don't inadvertently use the wrong social mores with anyone"

Some experience a later change in gender as a deep betrayal, particularly if they considered the relationship to be intimate. One MUD participant said:

Back when I viewed MUDS as a REAL reality, I fell in love with a female character. . . . But anyways turned out "she" was a he. Since then my personal policy is to NEVER get involved with anyone on a mud in a deep personal way.

As these examples demonstrate, choosing a gender, even a neutral gender, doesn't free people from standard gender expectations. Such expectations also affect how participants are able to enact the gender identities they do choose. Consider, for instance, Amnesia's description at the beginning of this essay. Amnesia is the female character of a male participant. I had a long conversation with Amnesia about his/her portrayal of a woman character, some of which I've excerpted here:

You whisper "so I'm curious—if everyone knows you're not

female, why still the female pronouns? Continuity?" to Amnesia.
Amnesia whispers "'Amnesia' is a woman, and always has been.
Amnesia was (is) my 'ideal woman', and so is more caricaturial
than any real woman can be. I think that means her femininity
shows through easier via text."
You whisper "your 'ideal woman' is caricaturially female?" to
Amnesia.
Amnesia whispers "no, I mean that I have no real experience in
being a woman, so can only draw a crude image with a broad
brush when I'm acting."

Amnesia's belief that one can portray a female character
well only if one has been a woman reflects a cultural belief
that women and men are substantially different from each
other. Yet, by playing into these beliefs Amnesia was in fact
quite successful. According to both Amnesia and other participants on BlueSky, Amnesia successfully masqueraded as
female for over a year and only had to give up the guise when
he met several other MUDders in person. Note especially
that Amnesia's "jeune femme fatale" description, a weird juxtaposition of Lolita-esque pornography and Dungeons and
Dragons-type iconography, seems not to have impinged on
his ability to be convincingly female, despite its clear references to male-oriented literature.

Amnesia's successful masquerade thus probably relies
in part on the wishful thinking of the predominantly male
MUD participants. But it also depends upon the wider cultural beliefs that certain behaviors are feminine and certain
behaviors masculine. This puts all MUD participants in the
position of "masquerading" as their chosen gender, regardless of their "real life" gender identity. Everyone is in "drag"
on MUDs; being more or less female has no relationship to
one's gender identity off-line, as Amnesia and I discussed:

Amnesia whispers "when I was full-out a woman, the differential
was unbelievable and measurable."
You whisper "but you know, I haven't really noticed [being treated
different]. 'Course, I haven't been on here as a male, but comparing myself to other people, it doesn't really seem to me that I get
more attention. Heh. Maybe if I was male, I'd get *no* attention."
to Amnesia.
Amnesia whispers "you don't 'act female' in the traditional sense,
as far as I've seen."

> You whisper "ah. I suppose that's true. So maybe it's not females
> that get more attention, per se. Am I less a woman than
> Amnesia? ;)" to Amnesia.

To a certain extent, this is not significantly different from face-to-face interactions. We are all expected to perform certain roles and meet particular gender expectations in all of our social interactions. However, both subcultural norms and technical limitations of MUD communication can limit gender performance on MUDs to stereotypical caricatures. Conversations move quickly, and one's presence in the conversation is apparent only when one continues to speak. In the limited bandwidth of text, typed conversation is the only means of communicating gender identity, and communicating it in a complex or nuanced way can be very difficult.

Given that the stereotypical characteristics attributed to men are more valued than those attributed to women, it is not surprising that the limitations inherent in online gender portrayal work out a bit better for men than for women. As indicated above, I've made no special effort to portray myself as specifically female, merely designating my gender and including information indicating that I am female in my character description. This has worked acceptably for me, perhaps partly because my known role as researcher has deflected attention from my gender and partly because I'm not particularly "feminine" in my face-to-face interactions and therefore do not attempt to portray this type of personality online. But other women I've met have indicated that attempting to "just be themselves" online produces less than desirable results.

Several female BlueSky participants feel that their lack of computer knowledge in particular affects the type of identity they are able to enact. One infrequent female participant said she wasn't interested in all the "tech talk" that occurred on BlueSky, and that "they all consider me a bimbo here." Another more frequent participant, Sparkle, said:

> In the bar [on BlueSky], I most likely seem more flakey than I am,
> but that's mostly because they don't talk about things I know
> anything about. All I can do is crack jokes and laugh when I read
> something that's funny to me. ... For the longest time, I was too
> scared to talk to anyone on here, I just hung out in the bar and
> laughed - which is why people think I am a ditz

Masculinity similarly becomes a mask to be performed, but even extremes of "masculine" behavior such as obnoxiousness and bullying are less likely to reflect negatively on the performer than similarly extreme forms of "feminine" behavior such as coquetry or flattery. This is particularly true because, in the interwoven subcultures that make up the MUD social environment, intelligence is highly valued and attributed more often to males. Textual obnoxiousness can at least sometimes be read as the incisive application of a keen wit, whereas demureness or other feminine virtues are likely to come across as merely dumb. Sparkle's shyness and laughter earn her the reputation of stupidity (reported to me by several other BlueSky MUDders), while even the most notorious of bullies on BlueSky are credited with high intelligence and even appreciated for their humor value.

Conclusion

I don't present these examples of MUD behavior as typical—there are hundreds of MUDs on the net, and I spend time on only a few. What I've tried to show are some of the ways in which gender gets enacted online. What's important, I think, is that the social context on any given MUD significantly limits what kinds of identities can be enacted there. You can set whatever gender designation you want and describe yourself as you will, but if no one believes your presentation, it won't be effective. If no one likes the way you present yourself, or thinks you're dumb because you present yourself in a stereotypically feminine way, that too will be limiting. In this way, online socializing is little different from face-to-face socializing. Social expectations shape who we can be.

The stereotypes of masculine and feminine identity found on MUDs aren't new. Nor is the higher value placed on the "masculine" characteristics of intelligence and aggressiveness. But the greater male presence online and the limitations of this form of textual communication create a context in which these stereotypes are relied upon to a greater degree. So the answer to the question with which I began my research is that gender, in fact, has a great deal of meaning online. Although individuals can choose their gender representation, that does not seem to be creating a context in which gender is more fluid. Rather, gender identities themselves become even more rigidly understood. The ability to change one's gender identity online does not necessarily result in an understanding that gender identity is always a mask, always something merely performed. Rather, there can be an increased focus on the "true" identities behind the masks. Further, what I've found is that the standard expectations of masculinity and femininity are still being attached to these identities.

Given all of this, is there a reason to be concerned about women's relative lack of presence on MUDs? I think there is cause for concern. Like other social forums, MUDs can be a place for meeting people with similar interests and for exchanging useful information. I could list dozens of examples of people (men and women) who've found jobs, relationships, etc., through MUDs, and many, many more examples of the exchange of useful information. BlueSky in particular often functions as a forum in which to discuss computer programming and network administration. People frequently log on from work with work problems, which other participants help them solve. To the extent that MUDs remain a hostile environment for women, they may become similar to "men's clubs" in which networks are formed for the economic and social benefit of members, largely to the exclusion of women.

Also, as my own experience shows, merely adding more women might not affect the situation greatly. I trade objokes with the rest of them on BlueSky (although I've noticed that with one other exception, I'm the only woman who does so). While I'm less likely to participate in other forms of gross and blatantly sexual forms of humor, I rarely object to them. Like any newbie, I want to fit in, and the social context into which I must fit has been largely set by men. It is this social context that needs to change before MUDs can be truly beneficial spaces for both women and men.

This is the point at which I should suggest a possible solution, but solutions are, of course, more difficult to identify than problems. At the least, it is important to note that the online environment is not itself a solution. Understandings of gender and the hierarchical arrangements based on these understandings do not simply disappear in forums where we can't see each other. We carry these understandings with us and re-create them online. Therefore, the appearance of more women on MUDs, and online generally, is likely to help only if both women and men make specific efforts to counter the types of stereotypical understandings I have identified. Ultimately, I think the situation on MUDs, and online generally, is more a symptom of wider social problems than a cause in itself. The status of women, and understandings about gender, need to be addressed societywide. Otherwise women on MUDs are unlikely to be much more than the objects of objokes.

Endnotes

1. I have changed the names of all the MUDs and MUD characters mentioned in the essay. All quotations cited in this essay were gathered from logs of my participation on several MUDs, from postings to Usenet newsgroups relating to MUDs, and from an email list also relating to MUDs. All are quoted verbatim.

2. "Lag" refers to a delay between the typing of commands and the execution, caused by network delays, processing delays locally on the computer where the MUD is located, or the MUD itself being slow. A MUSH is a type of MUD program.

3. "'TinySex' is the act of performing MUD actions to imitate having sex with another character, usually consensually, sometimes with one hand on the keyboard, sometimes with two. Basically, it's speed-writing interactive erotica. Realize that the other party is not obligated to be anything like he/she says, and in fact may be playing a joke on you." From "Frequently Asked Questions: Basic Information About MUDs and MUDding," a document regularly posted to the Usenet newsgroup, *rec.games.mud.misc.*

Like Magic, Only Real

Tari Lin Fanderclai

Fall 1994

The English department's tiny computer lab was packed, and at least half of the machines were taken up by my first-year composition students, all of whom were typing furiously, lines scrolling rapidly up their screens, pausing now and then to consult a handout or to scribble in a notebook. But their more obvious behaviors were giggling, poking at their neighbors, pointing to each other's screens, waving papers at each other, and talking in the tones of people who are trying to be quiet but can't quite contain themselves. I cringed a little as one of the more serious and businesslike faculty members appeared in the doorway; as manager of the computer lab, it was part of my job to see that students used our limited resources to do work for their English classes, reserving their "goofing off" for the larger, better-equipped campus computer centers. Yet there I was at the front desk with my own students having what could easily be viewed as a free-for-all while other students stood in the hallway waiting for a turn at a computer.

As the professor hesitated in the doorway, regarding the scene dubiously, one of the noisier of the crew propelled his chair across the short distance from his workstation to my desk and flopped a printout of his paper in front of me. "Hey, Tari, one of my group members from California says I should change this part of my paper," he said, and began to describe the recommended revision, clearly pleased with the suggestion, showing me how it would work and why it would be good for his essay. It was also fairly obvious that he didn't need help—he just wanted to show me the new idea, and after a couple of minutes in which he talked and I nodded a lot, he rolled back to his computer and resumed his enthusiastic typing and scribbling.

I felt a little safer—my student had demonstrated that he was indeed working. I turned to the professor, who was now standing near my desk. I could almost see her struggling to reconcile the apparent chaos before her with my student's articulate discussion of revision. Too many of us who have been trained to teach have been trained to believe that these are truths: First, that learning can be fun and a teacher should strive to make it so; and, second, that students who are having fun are not on task. And somehow the class activities we design with the first truth in mind are tempered by our belief in the second; as a result, our lessons generally hover somewhere around "more enjoyable than a dental checkup" on the fun scale. Rarely do we have to confront evidence that not only can learning be fun—it can even involve behaviors that our training tells us signal plain old goofing off. Such scenes are difficult to judge accurately—even if one has intentionally created them, the impulse is to shout for order, if only to relieve one's own cognitive dissonance.

Nervously, I began to explain: "They're using a site on the Internet to talk in real time to members of a writing class from California. We all watched the same movie and discussed it with each other online and wrote about it and then they emailed their papers to each other and now they're discussing revisions with their group members..." I trailed off awkwardly and went to work on the damaged disk the professor had brought with her, salvaging her files, and, I hoped, some of my reputation. In the back of my mind was the thought that maybe she'd be so relieved that her work was safe she'd forget to report to the director of composition that I was down in the computer lab encouraging bedlam.

For the chaos was necessary. My students were MUDding.

Spring 1993: The Beginning

As usual, I had stayed far too late in the computer lab, where I taught writing, administrated the network, and managed miscellaneous daily operations of the lab. I had gone in a short time from technophobe to computer addict, and I'd taken to staying long after the lab was closed. Free from users with questions and printers with paper jams, I could read manuals and tweak the network and subscribe to way too many electronic discussion lists. Recently a posting on one of those discussion lists had mentioned MediaMOO,[1] which I later discovered was a MUD operated by Amy Bruckman at Massachusetts Institute of Technology as a space for media researchers to meet and talk and collaborate with geographically distant colleagues. At the time, I knew only that a few members of a discussion list for writing teachers who use computers in their classrooms had discovered a place on the Internet where they could talk to each other in real time.

Curious, I entered the telnet address for MediaMOO and followed the directions on the welcome screen to connect to a guest character. Just like that, I was in the Lego Closet, wearing the suit of the Violet_Guest. The description of the Lego Closet reminded me of the descriptions of rooms in the text-based adventure games I was fond of, and with no better clues to go on, I tried a few of the commands from those games, pleased when they worked.

I discovered how to look around, move from room to room, and consult the online help. Shortly, I wandered into a room called the TechnoRhetorician's Bar and Grill. The room contained, among other things, a buzzword generator that when cranked emitted strings of buzzwords in nearly sensible order; a bartender who claimed to make good coffee; a boat you could really board; and several people whose names I recognized from the discussion list where I'd gotten MediaMOO's address.

I joined the conversation, a continuation of a topic that had come up that day in an asynchronous electronic forum most of us were using in conjunction with the annual Computers and Writing Conference taking place that week. Typical "newbie," I also examined and played with everything in the room. I cranked the generator over and over. I ordered all manner of things from the bartender just to see the ordered item appear in the list of things I was "carrying." I wrote on a bulletin board and petted a dog. I followed a few other characters on a tour of various areas of the MOO, fascinated with the various rooms and objects and their descriptions. I found out where to download everything from basic instruction sheets, to the MOO programming manual, to a client that would improve my interface with the MOO. After nearly two hours, I disconnected reluctantly.

The following day I obtained my own character for MediaMOO. Over the next few weeks, I obtained characters at other MUDs, and learned to build and program simple rooms and objects. And I kept returning to places like the TechnoRhetorician's Bar and Grill—spaces where those with similar interests tended to gather. For, though the virtual environment and the rooms and objects were fascinating, the main attraction for me was the ability to talk in real time to people who shared my interests. In "real life," I was at the time somewhat isolated: My department had just acquired its computer lab and network, and the school had gained Internet access relatively recently. While I was fascinated with the relationship between computers and writing instruction, as well as with the computers themselves and the resources accessible through my Internet account, most of my colleagues had other interests entirely. Many lacked the time or the inclination to learn about computers and the Internet; many were self-proclaimed technophobes and Luddites, some of them frankly alarmed at the presence of so many computers in their midst.

Though a fair number of instructors soon became competent users and made innovative uses of the capabilities of the local network in the writing classes they taught, not even a handful of them understood what I was up to alone in the lab all those nights, and fewer still ever became interested enough to join me. I soon learned not to talk much about my work or my discoveries. Trying to explain something I was excited about to someone whose eyes were glazing over was too disappointing. I often missed the sense of community I'd felt when doing work that my colleagues were interested or involved in.

Asynchronous electronic forums—discussion lists, newsgroups, bulletin boards associated with professional conferences, and so on—relieved some of my isolation, but a few key factors kept most of those forums, especially those populated by academics, from providing me with more than a superficial sense of belonging. Posters in such forums generally sign their names, with titles and affiliations—certainly a legitimate practice, but one that often helps to draw lines between the "big people" and the "little people" using the forum. It is sometimes painfully clear whose posts are to be valued and whose are to be ignored. The exchange is more like a correspondence than a conversation—members read and post in isolation, with responses to a post arriving hours or days apart. Even where members claim the forum is informal and spontaneous, the posts often seem to have been composed with the same care one might invest in an article (and some seem nearly as long!). And on many forums, any post regarded as off-topic or chatty or frivolous is immediately flamed by busy list members who object to having to sort through "nonsense" to find what they came for. None of these characteristics are bad; many of them are precisely the things that make the lists valuable. But participating in such forums merely made me feel more connected and up-to-date—it did not make me feel a part of a community.

On a number of MUDs, however, I've found that sense of community. As with asynchronous forums, I am connected to people who share my interests, but MUDs provide something more. For example, the combination of real-time interaction and the permanent rooms, characters, and objects contributes to a sense of being in a shared space with friends and colleagues. The custom of using one's first name or a fantasy name for one's MUD persona puts the inhabitants of a MUD on a more equal footing than generally exists in a forum where names are accompanied by titles and affiliations. The novelty and playfulness inherent in the environment blur the distinctions between work and play, encouraging a freedom that is often more productive and more enjoyable than the more formal exchange of other forums. It is perhaps something like running into your colleagues in the hallway or sitting with them in a cafe; away from the formal meeting rooms and offices and lecture halls, you're free to relax and joke and exchange half-finished theories, building freely on each other's ideas until something new is born. Like the informal settings and interactions of those real-life hallways and coffee shops, MUDs provide a sense of belonging to a community and encourage collaboration among participants, closing geographical distances among potential colleagues and collaborators who might otherwise have never met.

The Netoric Project

As members of the computers and writing community continued to use MediaMOO and to realize the value of the space to our own work and to our field in general, we felt the need to draw more of our colleagues in. One evening a few weeks after my first connection, I found myself on MediaMOO talking with Greg Siering of Ball State University about the possibilities of using MediaMOO for online conferences, and by the time I logged off, we had already begun to organize what eventually became known as the Netoric Project. With the help of several other MediaMOOers who were interested in computers and writing, we organized our first online conference, which took place on June 3, 1993. At that first conference we discussed the possibilities of MUDs for education and for professional collaboration.

Since then, the Netoric Project has expanded greatly; in the Netoric complex on MediaMOO we host weekly discussions as well as special events, and although our topics are chosen primarily for their interest to teachers who use computers in writing instruction, many of those topics have much broader appeal, and our events are attended by members of a variety of disciplines and professions from all over the world. A fair number of people in the computers and writing community point to the Netoric Project as the thing that got them started using MUDs, or hooked them up with distant colleagues to work on new research projects, or simply helped them feel more a part of a community that shared their interests.

At that first conference, we came up with a long list of

the effects we thought using a MUD might have on our writing classes. Students could collaborate with classes from other places, we said. They'd be exposed to new cultures and new ideas. They could try on new personae and test the effects of their words and ideas on immediate audiences, and they would learn all sorts of communication skills they could apply to other forms of speaking and writing. They would have constant access to their classmates as well as to others who shared their interests. Those who hesitated to speak up in class might be more willing to do so in the more anonymous environment of a MUD. No one, not even the teacher, could hog the floor; the power structure of the typical classroom would be disrupted. Our list went on and on.

And I was as enthusiastic as the next person about the list and believed a lot of the speculation would prove correct. But I suspect I wasn't the only one who was also excited simply by the notion of dragging a captive audience into one of those magical communities.

So I began poking about the net for a MUD that would allow classes. A few months later nothing would be easier to find than an education-oriented MUD, but at the time, I couldn't even find anyone who'd been to a MUD that hosted classes, let alone someone who'd used a MUD with a class. Finally I found WriteMUSH, operated by Jim Strickland and Marcia Bednarcyk. WriteMUSH, conceived as a place for writers to collaborate, had been up a few months and was prepared to host classes that wanted to collaborate on writing projects. I contacted Nikki Costello at Sonoma State University and Paul Bowers at Buena Vista College, both of whom I had met on MediaMOO. Both had helped in various ways with the Netoric Project, and, like me, they were eager to offer students the same kinds of experiences we'd found valuable in our own work. I arranged for each of my fall 1993 writing classes at the University of Louisville to work on a project with one of their classes. We spent the next couple of months planning, making arrangements with the WriteMUSH wizards, Jim and Marcia, and making our various individual preparations at our own sites.

. . .

Fall 1993: MUDding in Class

Nikki, Paul, and I had agreed that before our classes met each other we'd give them several kinds of preparation: They'd need to know the basic commands for navigating and communicating on WriteMUSH. They'd need to be familiar with basic netiquette and the particular conventions of WriteMUSH. And they'd need time to get comfortable in the MUD before we asked them to work together. So, about the third week of the fall term, my classes made their first trip to WriteMUSH. We had read WriteMUSH's policies, and I'd explained the general etiquette of the place, but I knew they'd just have to get in there and look around before any of it made sense. I gave them a handout on how to get connected, get their characters described, and walk around and talk to each other, and then let them proceed at their own pace while I walked around the room helping. It was, of course, chaotic: Some students were frustrated, others were unimpressed; some seemed mildly interested, and still others stayed long after class. By the time the lab closed that day I'd already talked with a student who was certain he couldn't stand this "MUSH thing," and I'd printed sections from the MUSH programming manual for two students who wanted to learn to make their own objects (something I didn't require, but certainly encouraged among those who were interested).

We spent some more class time just practicing and playing and talking through students' questions about MUSH commands, social conventions, and so on. At this point, they did not have a formal assignment; as learning to use a MUD requires quite a bit of problem solving, I simply relied on the MUD to provide us with whatever lessons we needed. After a few days, we went on to work more specifically on the possibilities and problems of communication and discussion in a MUD. I saved the logs of our first few online discussions and put them in a shared directory on our network; we used them not only to remember what we'd come up with in the discussions, but also to examine what happens when facial expressions, tone of voice, gestures, and other elements of physical presence are removed from a conversation. Already many students were drawing conclusions about effective communication, and, as we worked, we developed strategies for keeping online discussions focused, making sure we understood each other, and so on. Many of the students found they enjoyed the free-flowing, rapid exchange of ideas on the MUD, and within a few days they were so good at online discussion that we had to break into small groups using different rooms on the MUD to keep the conversation from scrolling by so fast that no one could read it.

Just using the MUD among ourselves proved a valuable complement to our real-life discussions, for the altered environment and the comfort of being somewhat anonymous (I had invited them to choose whether to use their own names and whether to identify their characters to the rest of the class) encouraged a number of students who were silent in face-to-face discussions to contribute to the online conversations. Some of the students who seemed more willing to talk on WriteMUSH commented to me privately that when they talked online, they felt as if they were being listened to according to what they said rather than according to how they looked, how they sounded, whether they were male or female or black or white and so on. For example, two students who were not native English speakers remarked that they liked having a place to communicate where they didn't have accents; a few female students said they felt they were taken more seriously online; and a student athlete told me he was less afraid of being taken for a "dumb jock" when those he was talking to couldn't see him.

Not everyone was enjoying the experience, of course. Some said they felt silly typing to people who were right there in the room. Others complained that the MUD environment favored fast typists. But I pointed out that soon we'd be talking to people we couldn't talk to any other way, and I offered them a shareware typing tutor, and they agreed to humor me a little longer.

By this time, we were beginning to think about two categories of issues: the usefulness of MUDs as work spaces, and the differences between meeting people and communicating with them online and in person. So, for our first real project, I gave them a chance to explore those further. I invited my Netoric Project partner, Greg Siering, to meet with my students online. I told them a bit about Greg and the work we'd done together, and I asked them to prepare questions for him both about how he felt MUDs were important to his work and about the differences between MUD communication and other kinds of writing and talking. Greg told them about several projects he was working on with people he hadn't yet met except online, recounted some of his experiences with communication and miscommunication, explained some of the conventions MUDders use for making themselves clear (for example, typing :-) to represent a smiling face so that others know you're kidding), and recounted some of his experiences meeting people he already knew well online.

Following their session with Greg, I asked the students to start drafting a paper on what learning to communicate on WriteMUSH had brought to their attention about other, more familiar kinds of communication. Their discoveries included realizing how dependent they were on giving and receiving physical cues during a conversation, how disconcerted they were when they didn't know the age or gender of the person they were talking to, and how seemingly little things like punctuation can make a big difference in whether your audience understands what you intended.

A few days later, after they'd had a chance to think about what Greg had discussed with them and about the perceptions they were forming themselves, Greg drove down from Ball State to Louisville and joined my classes in person. As he continued discussing their questions with them, they were able to compare the online and in-person experiences as well as think about the differences between a person's online voice and persona and that person's real-life presence. Later I asked them what they'd expected Greg to be like. Several had thought he'd be older than he was, and pointed to places in the log of our online discussion where the way Greg had phrased his points made them think of "some old college professor with a pipe and one of those jackets with elbow patches." One student confessed that he'd pictured "a big fat greasy guy" because Greg had mentioned he was eating junk food while talking to us. By getting them to talk about such seemingly trivial misperceptions, I got the students to think about how much our communications with others are influenced by factors about that person's physical presence—so much so that when we don't have those details, we might be tempted to make them up on the basis of very little evidence.

The students' experiences talking with Greg online and in class proved valuable in several ways: They came to understand more about why we were MUDding in the first place; they were able to think more concretely about factors that influence communication and make it effective or ineffective; and they saw how a person's written words affect how that person is perceived—a point that both made them aware of the power of words to create, and gave them a healthy skepticism about what others created with words. I was impressed with the papers they wrote as a result of their initial experiences using WriteMUSH. Not only did they discuss in detail their observations about voice, tone, style, persona, context, and so on; many of them were also able to articulate ways they could apply what they'd learned to their writing and other kinds of communication. I was eager to see what would happen when we began to work with other classes.

We had already begun to exchange email with our distant classmates, introducing ourselves and working out discussion groups. One of my classes formed small groups with members of Nikki's composition class, and the other with members of Paul's introductory film class. The members of each combined class viewed a film that we would talk and write about. Nikki and I had chosen *House of Games*, and the class working with Paul's group watched *Roger and Me*. The members of each discussion group had three tasks: to use WriteMUSH and email to discuss the movie, to prepare a discussion of a topic assigned to their group and give a short oral presentation to the members of their respective real-life classes, and to exchange drafts of their papers by email and use WriteMUSH and email to give each other feedback on those drafts. We explained their goals and deadlines, but after making some suggestions about possible ways of getting things done, we left it to individual groups to work out their methods and to schedule their meetings on their own time. They could, of course, ask us for help, but we required only that each group report to us periodically on the progress it was making. This plan gave students more control over and responsibility for the collaboration, and let them experience the convenience and challenges of collaborating online.

We'd expected to have to prod people to keep up with their groups, and we did, occasionally. But mostly they prodded themselves and each other; most students took seriously the responsibility to keep up their individual contributions to their groups. Nikki, Paul, and I watched in frank amazement as students made plans, worked out schedules and solved problems; late one afternoon I logged onto MediaMOO to meet with Nikki and found her so pleased that I was sure she was actually hopping around in delight. Several groups of our students had had problems that morning—WriteMUSH had been inaccessible for part of the morning, and the students also had difficulties getting email to each other (they'd made some errors using their respective mail systems, and already disconcerted by the difficulties getting to WriteMUSH, they'd had a hard time figuring out what went wrong). But instead of walking away empty-handed and blaming their problems on technical difficulties, they'd stuck it out, doing what work they could with their on-site group members and poking at the mail and the MUD periodically until they'd gotten in contact with their virtual classmates.

Nikki and I discussed what might have contributed to their willingness to solve their problems instead of taking the easy way out. Most students seemed to be enjoying MUDding and working together, and that was no doubt a motivator. In addition, that the project trusted them to work out their own methods and schedules for meeting goals and that they were responsible to people outside their own on-site classes seemed to contribute to their willingness to work out the difficulties. Given the environment we were using, there was no way for a teacher to check up on or impose control over the work they were doing; they knew that all we could do was wait for their progress reports and their products. I think that knowing they had a lot of freedom and the responsibility to use it wisely and efficiently caused them to take charge in a way they might not have in a situation where the teacher could stand over them and watch as they worked.

Students seemed to transfer those "taking charge" skills to other work in the class as well. The kinds of daily plans I was used to making were increasingly unnecessary for those classes—they much preferred to know what their goals were, what they were supposed to produce, and when they had to have it done, and to tell me what they needed to do on a day-to-day basis. They seemed far less likely than most students in my experience to bring excuses rather than work to class; if they encountered a problem in one method of getting a job done, they used another.

As I'd hoped, students in both of my classes reported that discussion with people outside their on-site classes helped them to see the film they'd watched from more than one perspective and to strengthen their own interpretations. Small classes often develop their own identities as communities, a desirable condition for a writing class, since students become more comfortable sharing their writing and working on collaborative assignments. But, like any community, the class members can also develop a shared general outlook, thus seeing only a small set of possible views on a given issue and coming rather quickly to what they consider the end of the discussion. Often they agree so thoroughly on their points that by the time they're asked to write about them, they complain there isn't anything to say that everyone doesn't already "know." Having their views expanded, challenged, and sometimes changed by their virtual classmates meant they were not so quickly satisfied (and bored) with their initial answers and solutions, and many of them carried the habit of looking at issues from unfamiliar angles with them throughout the rest of the term.

The class that worked with Paul Bowers' film class got an especially good example of the usefulness of being able to exchange ideas and information. Paul and his students gave us an introduction to looking at a film from a film studies point of view—something I couldn't give them myself. When it came time to write papers, my students got their chance to be the experts, taking charge of the draft exchange, giving useful advice about writing to their virtual classmates and showing them how to make useful comments on others' writing. Paul and I both felt that the quality of thinking and writing in the *Roger and Me* papers was in many cases exceptional, and the students attributed some of their success to the opportunities they'd had to benefit from each other's areas of expertise.

Paul, Nikki, and I found that many of our students began to think of WriteMUSH as both a playground and a resource. I often found students online building and programming, activities that I think enhanced their problem-solving skills and their willingness to be self-sufficient. Indeed, I occasionally found myself biting my lip to keep from laughing as one student would ask a question and another would say, "Look it up in the help...think I'm gonna spoon-feed you?" Other times, I found students on WriteMUSH deep in discussion about a topic or assignment from one of their other classes, getting ideas and input from others. As one student put it, "It helped to kick the movie around with people from outside, so I thought I'd try it for this history paper."

Figuring Out What We Learned

Of course, not everyone was thrilled about using WriteMUSH, and not everything went smoothly. For Nikki and Paul and me, one of the most fascinating and difficult parts of the project came when we tried to evaluate it. We all agreed that we'd met many of the goals we'd begun with and that many of our students had learned more than we'd hoped. But actually pinpointing successes and failures proved almost impossible. Each time we started to say, "Well, here's a part of the plan that worked," we reminded each other that that part of the plan had undergone many spontaneous changes as our students negotiated with us and with each other; each time we started to say, "Here's a part of the plan that bombed," we reminded each other of something the students had learned as they worked through the problems created by some shortsightedness in the original plan.

I think now that the main reason we had a hard time deciding whether or not those collaborations were successful is simply that we were working in such an unfamiliar way in such an unfamiliar environment. We'd placed students in a situation where they were in charge of their learning, putting in their hands many of the decisions about how to get to their goals. We'd encouraged them to build the same kind of community we felt ourselves a part of: a community of people working and learning and negotiating on an equal footing with each other. Our experiences with those communities fortunately held us back during the project when our conventional notions about what teachers are supposed to do threatened to control our actions; when one of us wanted to plan too rigidly or to interfere with a group that wasn't doing things the way we would do them, we reminded each other of the chaotic nature of our own online meetings and collaborations—the silliness, the arguments on the way to compromise, the fits and starts, the odd and unconventional ways we sometimes got from Point A to Point B. Yet for all that, when we tried to evaluate what we'd done, we found ourselves again using our old ways of thinking: We were trying to evaluate the project in terms of what we'd taught our students.

But the truth was that we hadn't taught them anything. Certainly we'd set initial goals and made the initial plan. But the students made the decisions about how to meet the goals, and they made many changes in the original plans; and as they worked they taught themselves and each other. To evaluate the experience, we had to think about what students had learned and how the entire situation, created by us and by the MUD environment, and modified by the students as they worked, had contributed to what they'd learned. The most important things we learned to do as teachers were to set clear goals, give students a variety of tools and skills to use in meeting those goals, and get out of the way while they worked through the problems on their own. Plans that didn't work and other difficulties that arose did not necessarily signify failures; when we let the students change the plans as needed and solve the problems in ways that worked for them, those would-be failures became useful learning experiences.

And thus I discovered one of the biggest values of MUDs to education. Many of us would like to create in our classrooms an environment where people can work the way that suits them best, where students have power and freedom and responsibility, where students actually take charge of their learning and develop skills they see the usefulness of and can transfer to other areas. But most teachers and students bring baggage to class that's difficult to get rid of, especially if the class is in the typical classroom at a typical school. For example, a common way of attempting to reduce the perception of teacher as authority and to put class members on equal footing is to arrange the chairs in a circle or around a table. Conventions die hard, though, and more often than not, everyone ends up directing most of their remarks to the teacher, who ends up asking most of the questions and prodding the discussion along. Given choices about how to proceed toward some goal, students often do nothing. They aren't used to making their own decisions, and they know that if they make none, the teacher will take over.

The MUD environment, however, can provide a place to shake many of the conventions of traditional education. As

Nikki and Paul and I discovered, when students have a list of goals and a variety of tools they can use to meet them, and when they are placed in an environment the teacher really cannot control, the rewards are many—and sometimes unexpected.

It still isn't easy; placed in a new environment, our tendency is to look for—or create—what is familiar and comfortable, often without stopping to consider what benefits the new environment might offer as is. Certainly a MUD room can be programmed to simulate a real life classroom; in fact, in some ways it can control students better than a traditional classroom. In a face-to-face encounter, a teacher can only ask for cooperation; in a carefully programmed MUD room, the teacher can enforce, say, a request for silence by typing a command that makes it impossible for anyone to speak without being called on. Such MUD environments exist, and they perpetuate many of the other conventions and limitations of traditional education as well. You can find faithful reproductions of the typical university on any number of MUDs, complete with separate buildings for each discipline, offices for teachers, and various disciplinary bodies that spend most of their time thinking up horrible things students might do and then writing elaborate programs to stop them from doing those things (and never mind that no student thought of such a behavior before hearing there was a specific program written to prevent or punish it).

It seems to me that such MUDs cater mostly to those who want to affect innovation and novelty while maintaining the traditional authority and control of the teacher. Indeed, it is always easier to introduce a new gadget than to adopt a new philosophy. I submit, however, that little is to be gained by taking students to an online representation of a conventional classroom. I would instead encourage teachers to look at MUDs as they are: full of creativity and cleverness, playfulness and magic; brimming with multi-threaded, free-flowing conversations and interesting people with various areas of expertise, and chock-full of opportunities to collaborate and invent problems and solve them. And I would ask teachers to think of ways to exploit those inherent characteristics of MUDs—not to try to override or control them. Given the chance, students make the most of alternative learning opportunities. All teachers need to do is get out of the way more often.

Forward

Since that first online classroom experience, I've brought a number of classes to various MUDs for a variety of purposes. All have been educational experiences; many have been great fun, and some have quite frankly flopped. Often I've been pleased by the students' accomplishments, a number of which went far beyond my expectations. The Netoric Project continues, and we've hosted many successful events. I'm administrating an educational MUD where I've been given the resources and freedom to put my ideas about educational MUDding into practice. MUDding is a part of my daily work and play.

In the last couple of years I've watched the mushrooming of educational MUDs, and I've engaged in heated arguments about what kinds of educational activities MUDs are and aren't good for; indeed, I have ranted repeatedly about the uselessness of delivering a lecture in what is designed to be an interactive environment, and the irony of programs that silence people in an environment that can let everyone talk and be heard at the same time, and the silliness of separating disciplines into different buildings in an environment that could blur traditional distinctions and let us learn from all those various perspectives, and...well, the list goes on and on.

But I'm always ranting about the same thing. Near the end of the term, in one of those very first classes I introduced to MUDding, one of the students wrote, "It's like magic, only real, and you can figure it out and learn stuff from it that you weren't even trying to get." Later, as we talked about her paper, she pointed to that sentence and remarked, "Ugh, that's awful."

"I dunno," I said. "I like it. I think I want to remember it." She looked puzzled, then less so; then she laughed, bracketed the sentence, and handed me that page from her printout.

I still have it.

Endnote

· · · · · · · · · ·

1. MediaMOO is a MUD; MUD stands for "Multi-User Domain."
MediaMOO is a MOO, which is a kind of MUD; MOO stands for
"MUD, Object-Oriented." Other kinds of MUDs include MUSH,
MUCK and MUSE. Although there are differences among the various kinds of MUDs, the basic idea is the same: Users connect to the
MUD using telnet or any of various MUD clients. Each user con-
nects to his or her own character (some MUDs provide guest characters for users who have not yet obtained characters of their own).
Using those characters, users can move around the various rooms
on a MUD, talk to other users, and interact with the various objects
on the MUD. Everything a user sees on a MUD is presented in text;
for example, when the user moves his or her character into a room
on a MUD, he or she is presented with a text description of the room.
On many MUDs, users are able to add their own rooms and objects
to the MUD.

Coming Apart at the Seams:
Sex, Text and the Virtual Body

Shannon McRae

In the decade since William Gibson coined the term
"cyberspace,"[1] the "consensual hallucination" he envisioned
has become a part of reality. Like the national highway system to which they have been all too frequently compared,
computer networks have drastically reconfigured the cultural
landscape. Gibson's paranoid vision of a world rendered
nearly uninhabitable by multinational corporations, whose
hegemony is enabled by means of a vast, interlinked information network, started out, like most good science fiction,
as social criticism. Now it has become the model upon which
various corporations, keeping up with enormous consumer
demand, are carefully planning and busily constructing a
brave new world.

In Gibson's novels, the technology from which cyberspace is accessed is available only to an elite: the corporations that buy, sell, and jealously guard information and the

highly skilled "net cowboys"—hackers who make a game and an art of stealing it. In reality, cyberspace is accessible to anybody with a PC and a modem. For-pay online services such as AOL, CompuServe and Prodigy, as well as numerous local and national bulletin board systems (BBSs), have proven immensely popular, with a user base that has expanded exponentially within the past few years. The hackers, computer scientists, academics and researchers who once made up the resident population of the Internet have recently experienced a phenomenal population explosion of eager new settlers, due mostly to a combination of media exposure and the increased accessibility of the Internet. In response to public demand, the popular subscription online services have begun to provide Internet access to their members, other for-pay access providers have begun to proliferate, and easy-to-use graphical World Wide Web browsers such as Mosaic and Netscape have only added to the traffic.

Neither the remote, parallel universe of Gibson's fiction nor the scary, computer-generated universe of films such as *Lawnmower Man*, VR (virtual reality) has become not so much a fiction as a condition, an alternative way of relating with the world and with other human beings. Virtual existence has become so immediate that what constitutes reality has become a very complicated question.

In the collective imagination, the figure of the cyborg—part human, part machine—has become a significant icon of this new, computer-generated reality. Because technology has become so much a part of everyday life, feminist theorist Donna Haraway concludes that "we are [all] cyborgs."[2] Gleeful cyborgs, the illegitimate offspring of the founding fathers of military and cybernetic research, fulfilling one of childhood's most avid and most forbidden wishes, are out playing on the highway that their fathers built. And in any variety of makeshift, sprawling encampments alongside the highway, they discover other shape-shifters, gaze into other eyes, glowing, opaque and shining phosphor-bright. An entire hybrid generation has redefined the concept of "doing it in the road."

"Virtual sex" is a generic term for erotic interaction between individuals whose bodies may never touch, who may never even see one anothers' faces or exchange real names. The phenomenon can include phone sex, exchanges of email, and encounters on chatlines, BBSs, and other virtual communities on the Internet, which will be my focus here. Media discussion of virtual sex on the Internet, or "netsex," tends to classify it with the type of fleeting, anonymous erotic experience obtained in sex clubs or pick-up bars or by calling professional phone-sex lines, and describes the phenomenon as the result of technologically-mediated alienation motivated by fear of AIDS or of strangers, or of the body's rapidly increasing redundancy in an age of progressive denaturation.

Certainly virtual sex has come to the fore at the same point in history that various means of deliberately altering the human body—some of which had previously been confined to specific subcultures—have become increasingly popular: bodybuilding, dieting, working out, piercing, tattooing, plastic surgery. One way to interpret the increased popularity of these practices is that people's bodies are all they have left to control for themselves.

There is, however, a more positive way to regard it.

Demonstrating an adaptability admirably in keeping with the seemingly endless evolutionary permutations of capitalism, human beings have turned the machinery of power that surrounds them into sources of play and pleasure.

Since virtual sex has gotten so much media attention lately, I want to make clear from the outset that not everybody who socializes virtually is interested in virtual sex. It might even be fair to say that most people who go online are perfectly happy to chat, discuss their hobbies and interests, and simply make new friends. If you are interested in meeting new people, or experimenting with virtual sex, it is socially inadvisable to approach anybody directly. The rules for meeting people, making friends and maybe getting laid are basically the same as in "real life," and you could find yourself becoming instantly unpopular if you make the wrong assumptions.

It is also important to note that, despite several recent, sensationalist media stories, the actual incidence of sexual predation, pedophilia, exchange of child pornography and other illegal or dangerous behavior is extremely rare. The net is a very safe, very enjoyable space for women and, with parental supervision, is a wonderful place for young people to explore and make friends.

A casual survey taken among individuals who've engaged in some form of virtual sex might suggest that it's no more compelling than, as one person put it, "an interactive *Penthouse* Forum." A lot of people who try virtual sex for the first time find it to be disembodied, alienating and not in the least sexy. Others, however, have discovered that it can be as involving, intense, and transformative as the best kinds of embodied erotic encounters, and that, furthermore, its virtuality enhances rather than detracts from the experience.

There is a certain kind of freedom in virtual sex. You can look however you want and do whatever you want, even push limits that for whatever reason you might not want to push in real life. In virtual reality, mind and body, female and male, gay and straight, don't seem to be such natural oppositions anymore, or even natural categories to assign to people. The reason for this is simple: In virtual reality, you are whoever you say you are. And, as some people who thought that virtual reality was a fun place to be somebody else for a while have discovered, it's often true that who you are is a lot more complicated than you ever imagined.

Virtual reality can mean different things to different people. Commercial, military and academic technological developments, which include head-mounted displays, data gloves and certain interactive video games, have been regarded in the popular press since about 1989 as the present and future of virtual reality. Although this type of technology has been well-funded, academically theorized and vociferously hyped in pop-culture lifestyle publications such as *Mondo 2000* and *Wired*, projected as a dangerous, scary fantasy in mainstream movies such as *Lawnmower Man*, or else as a sci-fi wet dream involving sex in wired body suits, very little of it is developed beyond preliminary stages, affordable or available to the general public.

Another type of virtual reality involves much less high-tech hardware and is made up by its many users as they go along. This type of VR includes the user-created environments found on telephone and online chatlines, bulletin board systems, newsgroups, and any of the number of commercial, online service providers that have proliferated in the past several years. Of this type, the richest, most complex and most comprehensively elaborate environment are MUDs: multi-user dimensions (or dungeons).

MUDs are text-based, interactive databases located on the Internet that allow many users to communicate with each other at the same time by typing text into their computers. Everybody in the same "room" sees whatever you type, and you can see what everybody else is typing as well. You can look at other people and the room you're in, walk or teleport into another place, pick up objects, even dance and get drunk in a virtual club. The only technology required is a computer and a modem or a means of direct access to the Internet. The only skills required are patience enough to learn a few simple commands, some programming ability or willingness to learn, and good writing skills to craft highly complex, vivid environments in which you begin to feel as if you're actually there. Unlike the better-known chatline systems, in which users can only converse as if on a telephone, players on MUDs are also objects; they have bodies they can describe as simply, attractively or fantastically as the skill and imagination of the individual writer allow.

Their users, many of whom have been online for years, regard MUDs as communities that allow for very real social and emotional engagement, political activism and opportunities for collaborative work on various civic, technical and artistic projects. The ways that people spend their time depend on the particular MUD. Most are used primarily for Dungeons and Dragons-style gaming. MOOs[3] such as the famous LambdaMOO are used for socializing, academic research, conferences and experiments in community organization or collective programming. Other server variants (for instance, MUCKs, MUSHes and MUSEs) are often used for solo or collective gaming, socializing or sustained role-play, although they may also be used for more work-oriented pursuits. Some examples of role-playing include playing famous, fictional characters from science fiction or fantasy novels or playing animals or vampires. While people seriously involved in role-play rarely, if ever, break character, most people, after spending some time in these virtual worlds, end up being more or less themselves.

When you newly arrive on a MUD, your first act is to decide on a name for yourself. On some MUDs, people go by their real names, but on most, people invent new identities. You can be a character from your favorite book, movie or TV show, or invent a persona of your own. You describe yourself and choose a gender. You can "morph" (change from one description and/or gender to another) with a single command, and change your mind and rewrite yourself at any time. Communication with other players is just as easy. If you want to say something, you type *say* followed by whatever it is you want to say.

If you (with character Amalea) wanted to make a face or a gesture, you might type:

emote looks baffled and her lip quivers slightly.

Everyone in the room would then see:

Amalea looks baffled and her lip quivers slightly.

Emoting allows for a richness and variety of communicative nuances not easily conveyable in other electronically mediated environments. Players become conscious of having "bodies" and, just as they do in "real life," express themselves with physical gestures as often as they speak. Sometimes the sense of presence is so vivid that you feel as if you really are touching, smelling, tasting, seeing whatever is around you, in a complex interchange of experience between a physical and an imaginary body.

The lack of physical presence combined with the infinite malleability of bodies on MUDs means that sex on MUDs is quite different from and possibly more intense than sex in other kinds of virtual environments. While many people engage in the fairly limited standard rituals of singles cruising, other MUD users seek out erotic experiences that would be painful, difficult or simply impossible in real life. For example, the residents of one highly popular MUCK[4] describe themselves as anthropomorphized animals. Although newcomers to this world usually get an impression of cloying cuteness, those with patience to seek out the large and active sexual subculture soon discover a number of unique sexual practices invented by and specific to players on that particular MUCK.

According to Kanu, a regular player there, one of the

classic forms of animal sex is predator/prey sadomasochism (S/M): "The submissive partner is eaten at climax. I haven't tried that, but I'm told it is interesting." Species are apparently chosen according to a highly complex social code: "Bears and wolves are usually dominant. Foxes are sorta generally lecherous. Elves are sexless and annoyingly clever. Small animals are often very submissive." Kanu, who plays a young, human male character ("lowtech, simple, friendly"), finds that while some other players find him quite attractive, others "are quite zoophilic and don't like having sex with a human."

Kanu described some of the wilder types of animal play to me:

Some players sort of invent new kinds of sex organs. For example, there was a centaur that had a really HUGE cock. But he was submissive, and what he liked was for people to fuck his cock. It was like a vagina on the end of a giant penis.

When I ask how animal forms might enhance role-play, Kanu suggests that they might allow for more primal experimentation, including the sensual and psychic effects of "predation, claws, teeth, size and strength parameters." Also, animal players enjoy playing with various sizes:

Some players are small. Like an ermine who likes me always wants to be taken home and used like a sex toy. The ermine would climb up my pants leg, etc. I didn't really get into it. Too silly for me.

Men are not the only ones who enjoy rough animal play. Women especially have found MUDs to be places where they are freer to experiment with their sexuality in ways that aren't so easy in everyday life. Two women who in real life are neighbors and best friends became lovers in VR. Although they enjoy various types of erotic role-playing such as being men, vampires, favorite TV characters, or invented personalities, Dan finds her friend's jaguar form to be particularly exciting. "He savages me," said Dan. "NOT something I'd want done to me in real life. Wounds heal quickly in VR."

While animal sex is one of the more exotic forms of experimentation possible in VR, gender-changing is probably the most common. Many people have found that having the freedom to be a different gender for a while can have very unexpected results.

Reflective of the demographics of net users, most MUD players are young, heterosexual males between the ages of nineteen and twenty-five. A surprising number of these young men try, at one time or another, to pass as female. Their reasons vary. Stephen Shaviro, a professor of English who has spent time on various MOOs, is somewhat cynical about their motivations, or about what gender-switching might mean about human behavior:

But let's not get carried away with utopian fantasies. Most straight men are assholes, and the mere opportunity for expanded gender play on the Net doesn't do anything to change that. A successful drag performance is harder to pull off than you think. Straight guys often pretend to be girls on the Net—I've done it often myself—thinking that the disguise will make it easier to score with 'actual' girls. But what goes around comes around: The girls these guys meet usually turn out to be other guys in virtual disguise. Face it, the information of which most straight men are composed is monotonously self-referential: It just turns round and round forever in the selfsame loop.[5]

When I asked some players I knew to be men in real life why they enjoyed being women online, their responses were far more complex than Shaviro's assessment would allow. One of the more curious gender-bending practices that occurs is that a surprising number of men masquerade as women to seduce other men. While I've spoken to a few gay men who've tried this, more than a few straight men have become female for a while to experiment with other men. Some try it on a dare, perhaps as a characteristically nineties form of macho bravado, simply to see if they can pass, to be better women than women. A familiar rite of passage on LambdaMOO involves young men inciting other young men to write a convincing female description for themselves and then successfully seduce a player who is notorious for boasting of his netsexual prowess with women. For these boys, "passing" is a game, something to be gotten away with. Others are motivated by a sense of self-exploration, to see what sex is like from a different point of view.

It can easily be argued that the free experimentation with gender roles that MUDs allow only enforces preconceived ideas of gender, and to a certain extent that argument is true. To attract partners as "female," one must craft a description that falls within the realm of what is considered attractive, and most people do not stretch their imaginations much beyond the usual categorizations: Voluptuous breasts, slim waists, flowing hair and the like proliferate as quickly in VR as in any Barbie Doll factory.

Elaine, who in real life is male, considers herself straight and has never had a male lover in reality. After some experimentation on MOOs, she has found making love as a woman with someone playing male to be not only intensely erotic, but also a way to experience a side of herself she felt she hadn't been in touch with so easily in real life. For her, being female "has something to do with wanting to inhabit something—elan, humor, emotional presence, communication, words—that I felt so utterly close to but just had no male model for."

In another conversation, she described what it is like for her to experience herself as female and have sex with a man:

When you're getting fucked by a man there's this amazing thing...
you realize you're being given all this energy and power...it courses
through you and you can channel it, throw it back, turn
up the voltage, make it explode, shoot it out your fingertips...
Or just surf it like a wave...only it's both inside and outside you,
dissolving....

Elaine is well aware that her experience of sex is not the same as that of an actual woman, and she avoids making universal conclusions about gender: "God only knows what weird stuff I'm saying about femaleness and [maleness] and myself and who knows what, but I feel it...strongly." What she experiences as her female side seems to be part of herself that her socialization as a male somehow excluded.

One of the themes that seems to recur in stories about gender experimentation is the experience of different kinds of power. For Elaine, the most intense thing about being female was the experience of being on the receiving side of power, which she could experience inside herself and then exchange with her partner. Jel, who describes himself as "basically pretty shy," has different, but equally complex associations with femininity and power. He had been on LambdaMOO for nearly a year as a male, but found it difficult to meet people. As a casual experiment, he logged on as an anonymous Guest character one evening, set his gender to female and was astonished at the sudden attention: "Whee! Instant popularity!"

Like Elaine, Jel describes his experience as "a thrill, sort of like power. I could ignore or chat with whomever I wanted."

Jel's account of his experiences supports Shaviro's argument that playing female is an easier way for men to meet women. He found that he was making friends easily for the first time and that, as a woman, other women were much easier to befriend. The experience, for him, was at first strange: "I had to keep reminding myself I was a woman,... [and] I was never *really* sure if I was speaking to another woman....I also developed friendships/relationships with men, but, oddly, those didn't manage to last when I switched back to male."

Further discussion with Jel revealed that although he played female because he found it easier to have intimate discussions with women as another woman, his motivation wasn't simply "scoring." Jel found that being a woman around other women provided him with an experience of sexual and emotional safety that certain of his real life experiences as a man did not:

"You'd have to know that IRL [in real life] I don't get along with men. Most of my emotionally (as opposed to sexually) intimate relationships are with women."

Jel discovered that the anonymity of the MOO, as well as the opportunity to change gender, provided him with a place to explore personal issues around power and emotion. One of the ways for him to work out his own complex and sometimes problematic associations with masculinity and power was to experiment with being dominated as a woman by another woman.

Finally, Jel met Plastique, who self-defines as lesbian in real life, and is equally concerned with the relationship between gender and power. Because it involves so many of the interesting and complicated ways that gender and identity can become fluid concepts for people to explore, the tale of their relationship is perhaps one of my all-time favorite virtual love stories.

Like many other people, Plastique discovered that

LambdaMOO was an enjoyable place to experiment with sexuality. Although she occasionally played male, she usually played a version of herself. Her first virtual affair of this type, she explains, allowed "a feeling of power and agency with a man that I'd not experienced with straight men IRL....I thoroughly enjoyed the experience, especially because my male partners were so responsive to my sexual power."

Plastique first met Jel in his female persona. After an intense affair, she felt the need to break off the relationship:

The reason I think I fell for her and the reason I dumped her were the same. She did 'woman' in a way that fascinated and repelled me. She loved to be dominated...she wanted me to do that to her....I think I was repelled by what I saw in her and repelled that, on some level, I wanted that too.

Plastique then met a man online about a month afterwards:

I instantly fell for him when I met him. Hard. I knew it was trouble fast. I thought only fleetingly that they may be one and the same...both lived in the same city, both college students, both into the same perversities. Silly me. I just didn't want to find out cuz I thought if he ever met her IRL, they'd fall in love for sure cuz they were so alike.

The best part of the story is this:

He thought the entire time he was with me as his female character that _I_ was a man...he was convinced!....So he seduced me, not caring what my real life sex was. When he told me he was the same female character I'd dumped, I was both crushed, betrayed _and_ I fell even deeper in love.

Both Jel and Plastique discovered that VR gave them an opportunity to experience the various complex links between gender, sex and power in an environment that provided an opportunity to role-play safely and easily. Occupying another gender, playing out a sexual attraction that for one reason or another real life doesn't allow, or experimenting with roles involving dominance and submission are all ways to discover that sexual identity is much more complex than we allow ourselves to think.

For a woman, whose sexual power is often described in terms of her beauty or seductiveness, playing male can give her a very direct sense of strength normally associated with masculinity. Two women I spoke with describe a sense of bodily strength and also, curiously, an edge of violence they are not as aware of when they are female. As one woman put it:

I had no idea how much of my identity had been anchored in gender until I felt what it was like to fuck someone while I was a boy. It's like my virtual, male body and my actual female body aren't separate, but somehow doubled...I get this cock, and shoulders, and there was this sense of being much larger than my [female] lover, so there'd be this sense of having to be very careful not to overwhelm her, which was _intense_. With that, there's this almost violence that I feel like I'm always on the edge of, that's incredibly erotic.

Dan, who switches gender periodically, also describes distinctly different senses of embodiment:

When I'm female, I'm very aware of the male's strength, as I am IRL, and my own lightness, grace, whatever. One of [my female morph's] lovers is VERY big. A hunk. I don't go for that kind at all IRL, but on MOO, the threat isn't there...But my female character does _not_ want to be dominated. [My character] Dan likes it.

Again, the importance of these accounts is not that women can feel what it's "really like to be men," any more than the men I spoke with imagined that what they were feeling was authentically female. Rather, we can experience for ourselves, inside ourselves, the kinds of things that we associate with female or male, and realize that those aspects are not, after all, something Other and outside of us.

Dan, like Jel, finds that VR offers a certain sense of safety and security in pushing the limits of erotic experience: "[Dan's] maleness is definitely security. He's from a SciFi show...he came complete with a sadomasochistic edge I didn't invent....Dan likes to be dominated, but he puts on a hell of a fight." Another male character she plays, however, "tends to be dominant. It's intense being both of them....More so than with my woman characters, for MOOsex."

Dan describes the double-body sense that many players describe when they experience the intense but curiously dislocated eroticism of virtually enacting another gender:

When I respond sexually as one of my male characters, there are two very definite feelings of body involved. My own woman's body responds physically. But I'm not a "one-handed typist" during sex. When I emote that my cock is throbbing, I'm imagining that very vividly. Because body feels real in a sense in MOO, it has more sense of physicality than reading a story, or having a fantasy.

The term "one-handed typist" is a kind of joking way to describe masturbation during netsex. Some people do this regularly, others are so involved in the pleasure of their virtual character that they experience the described, virtual orgasm as if it were real.

Although many people find that the opportunity that MUDs provide for anonymous role-play gives them a special sense of freedom and power, some people discover that these same features make them far more vulnerable to deception and set them up for emotional pain they never anticipated. Because role-play involving dominance and submission can lead to an experience of intense vulnerability, experienced players know that it should always be engaged in within an environment in which both players feel safe. The anonymity, distance and partly imaginary nature of virtual space can lead to a situation in which one person feels deceived. As most people who spend any amount of time on MUDs discover, often to their surprise, the emotions involved when people become sexual with each other can be very, very real. Anonymity and distance can leave people feeling more vulnerable rather than less.

One woman I spoke with thought, like Plastique, that the MOO would be a "safer" environment to explore aspects of her sexuality she had never experienced in "real life": sex with men and playing out S/M fantasies. Because of her inexperience and the naivete that leads many new players to take others exactly at their word, however, Shade's first experiment was traumatic. When the boy she was playing with began to play more roughly than she felt comfortable with, she found herself far more deeply upset by the scene than her assumptions about VR's safety and distance had led her to anticipate. She abruptly left for her own room, but was too unsettled by the intensity of her experience to simply disconnect. A woman who had also been involved in the scene became concerned, paged her (paging is a means for players who are not in the same room to communicate with each other) and offered to ask a lesbian friend of hers to join Shade and talk her through her upset.

The friend, Trina, was very sympathetic and understanding. After a fairly short time, Shade became intimately involved with her. Their involvement included S/M role-playing, which she felt was much safer with a woman. When, after several weeks of intense sexual encounters, Trina confessed to being male in real life, Shade was furious. When she discovered that all of the people involved in the original scene knew each other well and were all playing along in the deception, her sense of betrayal was so deep that it was several weeks before she felt she could trust anybody she met virtually again.

According to Shade, the boy's rationale for passing as lesbian was, apparently, to "help women" and keep them safe from other men who might hurt them. I can't say what his actual intentions were, as he is no longer around to ask: The incident happened several years ago, and Trina has since either left LambdaMOO or, as sometimes occurs when someone becomes visibly unpopular, assumed another identity. Possibly this group of boys was simply experimenting with role-playing and had no idea their actions would be of any emotional consequence. For some participants, never seeing the face of whoever they're dealing with in VR somehow translates to not having to be responsible for their actions.

In any case, Shade's story is by no means the only incident of genuine emotional pain incurred in what newcomers naively believe to be an environment divorced from reality. Another woman I spoke to, who in real life is married and considers herself to be monogamous, found herself befriending and becoming emotionally close with someone she thought was another woman. Although they maintained their friendship when the character revealed his true gender, she felt their friendship had significantly changed. Rather than the experience of cooperation and mutual support she experienced with someone she thought was a girlfriend, she found that their dynamic changed to one of constant embattlement. This may have had as much to do with her anger at the deception as with the difficulty of maintaining cross-gender friendships. As she put it: "Sadly then, when it was a woman, it was 'help the sister' and she'll help you. When it was a man, it was the old, boring game of 'batter your head senselessly against a titanium shell.' Anyway, he changed my life."

That such masquerades can go undetected comes as a surprise to many, both on and off the net, who assume that fixed gendered behaviors exist and are readily detectable. Experience in VR indicates that this is not necessarily the case. Even people whose experiments with gender lead them to discover that it is not a fixed fact still find that each gender carries for them a set of assumptions, sometimes deeply unconscious ones, about what femaleness or maleness actually is. And people in VR find that their responses to whatever they imagine the gender of their partner to be are often more important than the physical actuality.

Explaining how it is entirely possible for otherwise highly intelligent individuals to be so deceived, one of my net friends, a professor of psychology, described for me the psychological concept of a schema:

Think of a farmer...got the picture?...In your head...of a farmer? Can you see the farmer?

[I respond that I get an image of Mr. Greenjeans.]

Good. that's a schema. I say a word... you see a whole set of things that go with it. Things that fit. overalls, flannel, tractor....Ok...so....I'm a woman. you have a vague image...very vague. But certain things fit within the limits....When I do something that confirms your impression....you suck it up without even thinking because...because schemas are our brain's way of being efficient. You only have to learn farmer once or twice. then you can just call it up. and you can add new stuff to it to expand. Like a guy in a broad straw hat bending over in a rice paddy, a woman in dry soil with a kid strapped on her back in africa.

But....mostly you call up greenjeans....So....our brains are trying to conserve tasks...can only handle so much at once. So we construct and call upon schemas for a great deal of stuff. one of our schemas is gender. We "know" what women are like. We're willing to be a little flexible but we don't re-invent our social construction of women every time we see one...we use our defaults and go from there. So when a character is female we bring the schema to bear on our impression of her. that's one concept. the other is about another cognitive process....

Ok...so the other part...why we aren't so good at seeing the [gender]bending: two reasons...one: there really aren't very many consistent differences in the behaviors of men/women. Not really. long story that i could explain with data but the differences are much much more situationally dependent than they are particular to a gender....Just because there is a significant statistical difference between men and women's behaviors doesn't mean that we can use that info to predict their behavior.

One way to avoid altogether the automatic assumptions people assign to a particular gender is to choose one of the alternative genders that some MUDs offer. The spivak gender available on MOOs, for instance, has a unique set of pronouns: *e, em, eir, eirs, eirself.*[6] It has encouraged some people to invent entirely new bodies and eroticize them in ways that render categories of female or male meaningless. Because the pronouns assigned to the participant efface gender distinctions, a spivak can have any morphological form and genital structure e devises for emself. Twine, a woman in real life, describes her experience of a spivak body with an imaginative richness and sensuality that pales typical accounts of sex involving male or female bodies:

For me, spivak is able to transform very quickly...well maybe gradually...say, grow a penis in a few minutes...? And two spivaks means that one could shape the other, as well....if the other allows the suggestions....and for me there's also these little extensions....like very fine root hairs on a tree root....anyway....these little hairs form lots and lots of connections....They are very sensitive....and as love making progresses....they stroke, penetrate, and even fuse. Also, spivak sex, for me, involved musical tones from deep inside the chest, much like cat's purring....and little chiming sounds from those tentacles.

The real-life genders of the spivaks I've spoken to seem to break down pretty much exactly fifty-fifty, female and male. Although their motivations for playing an indeterminate gender vary, most with whom I've spoken are making a conscious statement about the inadequacy of either female or male for expressing more complex aspects of human sexuality. The most articulate explanation I've seen was given as a recent posting on an anonymous mailing list on LambdaMOO:

The molds of MAN/WOMAN, MALE/FEMALE, MASCULINE

FEMININE do not fit everyone. They certainly don't fit my SO [significant other], who is genetically neither male nor female sexed, eir gender is equally "neither." E just doesn't fit into the "genders" described by fe/male. Neither does e want to. In this disembodied universe of MOO, where we can be whatever we choose to be, but must [wear] an artificial label of "gender," I think it's important to be able to select an appropriate one.

Also, I think I suspect there are a few card-carrying females out there who use the "other" genders to avoid the constant intrusions of [players] looking specifically for females. I know I've hidden under "other" genders for this purpose, and I was glad to have choices other than fe/male.

That the spivak gender allows people to construct their bodies in whatever way they choose foregrounds the fact that netsex is as much an act of writing as it is of sex. Unlike other erotic writing, which is static, netsex is interactive. The people involved are writing at the same time as they are making love. For Bret, the eroticism of netsex has as much to do with the play of language as the play of bodies. In the context of virtuality, gender becomes partly an abstraction— a feature of the particular bodies that are being written rather than an important fact of human identity. As Bret puts it, netsex "is more about sexy language than it is about the gender of my partner's body."

Bret, a woman that prefers to stay female in VR, prefers men but also avoids identifying her sexuality with specific terms. Her VR lover is a real-life man who tends to play female. When I asked her if she would characterize theirs as a lesbian experience, she responded:

It's not like we're playing at lesbians, if that makes sense....It's not a 'lesbian scene' at all. I don't think that there's any conscious decision behind it. We tend to talk about our bodies in terms of their specificity, so we'll talk about getting a couple of them together, but it will be because we want those particular bodies to fuck, rather than going for a lesbian/het experience.

For Bret, netsex is about writing the body. The erotic pleasures that it affords are as much from the experience of exchanging beautiful and arresting language as about describing what is happening to bodies.

Netsex allows two (or more) people to simultaneously write themselves and each other. The convention is that one person describes, in the third person, what she is doing to her partner, whom she addresses as you, for instance: "Amalea slips her fingers gently across your collarbones and kisses your mouth slowly, lingering, as if tasting something delicious for the very first time."

Amalea's partner might respond: "Jamie shivers, tongue slithering over yours, hands sliding down your back and pulling you hard toward him."

Not all netsex involves realistic descriptions of bodies engaged in specific actions, however. The fact that it is *written* sex allows people to be as graphic or as poetic as they choose, to take pleasure in the exchange of language with a lover. The following, written by Bret, is not an example of netsex, but rather of the poetic quality of language that some people employ:

....Joints poised lush, tumbling cats french-kissing in a sunbeam, heavy thumping counterpoint on the wild curves and the surprise of passing a stranger with a lonely chainsaw......

Movement bright and metallic flowing through ripples of grimy sure-muscled sheening sweat and a lick of dust ghosting jaw jutting angle cool and sure.

Stretch electric, bored shroud posing over flesh aware like pain, visceral skeleton grinding easy into the pretty red click of teeth on vertebrae tongue straining for hot marrow......

Another example of the kind of language that might be used is in this anonymous post from an erotic mailing list on LambdaMOO. While this, like the piece above, is more an individual fantasy than a two-person exchange, both are good examples of the literary quality of netsex:

I woke up too late in a slithery-soaked heat from a dream that must have been about you, because he touched me the way you used to.

Only, the body was different, not the delicate, strung wire of you, but a mass of muscle and bone between my thighs, sprawled apart like riding a horse.

I think it was this guy from one of my classes that I don't even care about.

I think it was you maybe, before I lost you.

You know the way dreams work, there wasn't a particular scenario to have led up to, only we were in a coffee shop maybe, and I had just ordered coffee and was sitting right next to you, my skin craving yours, couldn't move away, though I wasn't sure that you wanted to....

That I wanted to....

Only our arms were around each other then, and my palm sliding down your chest, your belly, fingers delighting, suddenly in the otter-smooth dark hair, and the groove of muscle below and the sweet shuddering of your skin. You pushed my hand down, suddenly, cock leaping into the curl of my fingers, pulsing smooth and warm and hard as sun-warmed stone, as memory and your fingers teasing sharp and burning sparks from my nipples, two fingers slipping slick into my mouth, suddenly, licksuck, burning trailing down to slide inside me but I want you now, I want you now......

My hands on your shoulders, pushing you back, down, scoop you up into me, falling on you in a long, burning slide. Not enough, not deep enough, you move, crashing under me, up into me....

I want to fuck your skin off.

I want to be inside the cage of your bones, want to swim in your pulse, I want to weave in and out between your ribs, I want to be inside you.

Your fingers bruising my hips, I want more, my hands braced on your shoulders, snapping down onto you, falling up, your eyes wide, astonished, biting your lip and my hair flying......

And I'm not sure what's bodies and what's liquid pulse anymore,

Or what is the burning edge of dreaming......

Waking up to cats, and tangled sheets and the heat of early summer and all these memories in my mouth and burning through my fingers.

A girl can dream.

To be involving, netsex involves a constant phasing, simultaneous awareness of the corporeal body at the keyboard, the emoting, speaking self on the screen, and the existence of another individual, real and projected, who is similarly engaged. Mind/body awareness is not split, but doubled, magnified, intermingled.

Virtual sex involves role-play. But when you are playing, the roles are entirely true. As in drama, ritual, liturgy and certain esoteric sexual practices, within the context of the scene being created, the play takes on such focus and intensity of purpose that the "I" becomes meaningless, standing outside the self, in a state of *ekstasis*, quite literally a being put out of its place, enraptured: seized by force, bursting, smitten.

When intense erotic union is accomplished in a kind of void, in which bodies are simultaneously acutely imagined, vividly felt and utterly absent, the resultant sense of seizure, of scattering, of self-loss can be experienced as a violent mingling of pleasure and pain. The intensity of pleasure results from the kind of sustained dislocation required when your body is entirely real and entirely imagined at the same time.

Because it involves the absence of actual bodies for its intensity, netsex is comparable to Courtly Love, the medieval European tradition in which very elaborate, sometimes extremely intense romances were played out between two people who were forbidden to be lovers. Usually, the male lover wrote impassioned poems to his beloved, or courted her with music or poetically eroticized conversation. It was important for the tradition that the beloved be unavailable, generally because she was already married, but sometimes for other reasons such as religious vows of celibacy. The fact that the physical bodies of the lovers could never meet only added to the intensity—the lover frequently attributed to his beloved highly idealized qualities and sometimes worshiped her as a goddess. Sometimes Courtly Love was expressed in religious terms. Likewise, certain kinds of religious devotions, such as the various accounts of saints, referred to God or Jesus in terms ordinarily reserved for lovers. The experience of sexual ecstasy and religious transport are, as many medieval saints knew, very similar. Sometimes netsex approaches the intensity of poetry or religion.

To sustain erotic pleasure while making love with words, lovers must maintain their powers of language at a moment when the power of coherent verbal expression is customarily abandoned. The speaking and experiencing selves are necessarily split by the requirement of maintaining language within sensation. Rather than a sensory void, however, the split can perhaps be described as a highly charged space, the delirious, lacerating edge of experience between the pleasure of the text and the point at which all language fails.

Paradoxically, the more intensely that individuals involve themselves in netsex, the better able they are to evoke bodily intensities in words, leaping into the gap between utterance and experience, simultaneously enacting the rush of bodily sensation and the writer's ecstasy at producing text, being-in-text and being-in-body.

Continually subjecting oneself to a condition in which bodily experience and emotion must constantly be expressed in words is not entirely unproblematic. Our culture has too frequently found the enormous range of emotions and experiences that are simply not expressible in language to be, therefore, unworthy of attention, to the detriment of real emotional health. Real, living bodies are already devalued in an age where the passive reception of selected information has come to replace lived experience as reality.

Nevertheless, sex, love, or pleasure in any form may well afford some measure of resistance against social and technological forces that would divide us from each other and prevent us from naming and shaping our own experience.

The machines that we once feared would overpower and control us, that we thought would make the world more inhuman, have become ways for us to experience intense pleasure with other people we might otherwise never have met. All the things that separate people, all the supposedly immutable facts of gender and geography, don't matter quite so much when we're all in the machine together. Eroticizing our technology might not mean giving up the ghost, but rather giving in to the pleasures of corporeality that render meaningless the arbitrary divisions of animal, spirit and machine.

Endnotes

1. William Gibson, *Neuromancer* (New York: Ace Books, 1984).

2. Donna J. Haraway, "A Cyborg Manifesto: Science, Technology,and Socialist Feminism in the Late Twentieth Century," in *Simians, Cyborgs and Women: The Reinvention of Nature* (New York: Routledge, Chapman & Hall, Inc., 1991), 151.

3. MOOs are MUDS, Object-Oriented, so called because they are programmed in an object-oriented programming language. Players can program objects that interact with each other and that can act as a basis on which to construct other, more complex objects.

4. MUCKs, MUSHes and MUSEs are all variations of the MUD

server, characterized by slightly different programming languages as well as varying degrees of programmability.

5. Steven Shaviro, "Doom Patrols," ftp://ftp.u.washington.edu/public/shaviro/doom.html.

6. The spivak gender category was originally invented by Michael Spivak in *The Joy of TEX: A Gourmet Guide to Typesetting with the AMS-TEX Macro Package* (Providence: American Mathematical Society, 1990).

Resources

An excellent source for electronic mailing lists is GENDER-RELATED ELECTRONIC FORUMS, compiled by Joan Korenman. Her site includes an extensive listing of publicly accessible electronic forums (or "lists") related to women or to gender issues. She can be reached at korenman@umbc2.umbc.edu. The entire list can be found at http://www unix.umbc.edu/~korenman/wmst/forums.html

An excellect source of electronic forums is Usenet, with its vast array of "newsgroups." Among the thousands of groups are: *soc.feminism, soc.women,* and *soc.men.* These newsgroups all carry discussions of male/female relations, as well as other topics. The newsgroups are public, open to both men and women. *Soc.feminism* is moderated; the others are not and tend to be somewhat wilder and more argumentative.

Women's World Wide Web sites include: Feminist Activist Resources on the Net: Communicating with Other Feminists, which can be found at http://www.igc.apc.org/women/feminist.html.

Another good guide is "Sources for Women's Studies/Feminist Information on the Internet." Laura Hunt's guide lists many of the mailing lists in addition to some explanations of Internet resources; it can be found at gopher://una.hh.lib.umich.edu:70/00/inetdirsstacks/women%3ahunt (about 73K). She can be contacted by email at lhunt@shadow.stjohns-nm.edu.

The Ada Project (TAP) is a WWW site designed to serve as a clearinghouse for information and resources related to women in computing. The goal of TAP is to provide a central location through which resources can be tapped on conferences, projects, discussion groups and organizations, fellowships and grants, notable women in computer science, and other electronically accessible information sites. TAP also maintains a substantive bibliography of references. TAP serves primarily as a collection of links to other online resources–rather than as an archive. http://www.cs.yale.edu/HTML/YALE/cs/Hyplans/tap/tap.html

Women Server, maintained by Jessie Stickgold-Sarah: http://www.mit.edu:8001/people/sorokin/women/index.html

Women's Resources Project: http://sunsite.unc.edu:80/cheryb/women/

Contributors

Judy Anderson "yduJ" has a B.A. in Philosophy and M.S. in Computer Science, both from Stanford University. Her first computer use ever was winter quarter of her freshman year, when she took an introductory programming class. In 1980 she began a part time job as system administrator at Hewlett-Packard and has been working at a variety of computer jobs ever since. She currently works for Harlequin, Inc. in Cambridge, MA. She can be reached at yduJ@cs.stanford.edu.

Paulina Borsook is a San Francisco writer who has a Master of Fine Arts from Columbia University. Her novella, "Love Over the Wires," was the first fiction published by *Wired* magazine. Her work has appeared in *Newsweek, Whole Earth Review, Newsday, Working Woman,* computer publications too numerous to count and literary magazines too obscure to mention.

Stephanie Brail has been hanging out on the Internet since 1988, when she started using email in college. Since then she's gotten into flame wars, hosted online forums, helped form a Usenet newsgroup about a very bad TV show and started a mailing list for women Web designers called Spiderwoman. In 1995, she founded Digital Amazon, a computer consulting business dedicated to empowering women through technology. She lives in Los Angeles with her rambunctious cat and four computers. As always, she is working on her first novel–to be finished sometime in the next millennium. She can be reached at pax@primenet.com.

L. Jean Camp works at Carnegie Mellon University in Pittsburgh, where she is getting her Ph.D. in telecommunications policy. She gets by there with a little help from her systers.

Susan J. Clerc is currently in her second year of a Ph.D. program in American Culture Studies at Bowling Green State University. Her master's thesis was titled "The Influence of Computer-Mediated Communication on Science Fiction Media Fandom." She recently completed a chapter for *The Truth is Out There: Reading the X Files* titled "Welcome to the World of High Technology: *X Files* Fans Online" from Syracuse University Press. She has a B.A. in History and Classics from SUNY Buffalo, a Masters of Library Science degree and a law degree. Her goal in life is to have more letters after her name than there are in it.

Karen Coyle has a Masters of Library Science degree and develops computer systems for libraries. She is active in Computer Professionals for Social Responsibility, a non-profit public interest group concerned with the impact of computers on society. She likes to take things apart and see how they work. In her first computer job, she was told "We're lucky to have you—if you were a man, we'd have to pay you twice as much." She now earns as much as a man, and works twice as hard. Her URL is http://stubbs.ucop.edu/~KEC

Michele Evard is a Ph.D. candidate working with Seymour Papert at the MIT Media Lab. Her research focuses on children's participation in online discussion groups. She has been a part of online communities related to both personal and professional interests since 1984, when she began pursuing bachelor's degrees in mathematics and computer science. She received her Master's degree in computer science in 1992 from the University of Oregon. She can be reached at mevard@media.mit.edu.

Tari Lin Fanderclai holds a Master of Arts in English from Western Illinois University and has taught college composition classes for ten years. She is one of the coordinators of the Netoric Project at MediaMOO and runs an educational MOO called Connections. She lives in Boston, where she does computer consulting. She can be reached at tari@ucet.ufl.edu.

Netta Gilboa is the publisher of *Gray Areas* magazine. Besides computers and hackers, she loves the art of conversation, cats, the music of bands like the Grateful Dead and Jefferson Airplane, watching movies and eating sushi. She can be reached at 76042.3624@compuserve.com or via snail mail through Gray Areas, Inc. at PO Box 808, Broomall, PA 19008. Her URL is http://w3.gti.net/grayarea/gray2.htm

Lori Kendall is a doctoral student in sociology at the University of California at Davis, where she researches issues regarding computer-mediated communication and culture. She is currently writing her dissertation on MUDs.

Shannon McRae is a Ph.D. candidate in the University of Washington Department of English. She has spent several years researching virtual communities, and writes from an insider's perspective. Her other interests include Medieval Irish myth and literature, Modernist poetry, gender and anarchism.

Judy Malloy has been making experimental books since 1975. Her early works used photos, text, and drawings on 3x5 cards in file drawers or in electromechanical books read by pushing buttons. She has had work published by E.P. Dutton, Tanam Press and St. Martin's Press. Her classic hyperfiction ITS NAME WAS PENELOPE is published by Eastgate Systems. She has a B.A. in English and Art from Middlebury College. She can be reached at jmalloy@ well.com.

Cathy Marshall is currently a member of the research faculty in the Department of Computer Science at Texas A&M University. From 1988 to 1995, she was a member of the research staff at Xerox Palo Alto Research Center. Cathy has led a series of projects investigating analytic work practices and collaborative hypertext, including two systems development projects, Aquanet (named after the hairspray) and VIKI. She can be reached at marshall@bush.cs.tamu.edu.

Donna Riley is a doctoral student in Engineering and Public Policy at Carnegie Mellon University and a founding member of the Clitoral Hoods. She can be reached at Riley+@ andrew.cmu.edu.

Laurel Sutton is a San Francisco Bay area linguist who works on issues of gender and computer-mediated communication. She is co-editor of the zine *Inquisitor*, available in fine book stores or at PO Box 132, New York, NY 10024-1032. She is also a founding member of Bitch Nation.

Ellen Ullman has worked as a software engineer and consultant for fifteen years. Her writing has appeared in computer industry publications, in a recent collection of essays from City Lights Books, and in *Harper's* magazine.

Lynn Cherny (right) is a researcher at AT&T studying multi-user synchronous communities on the Internet. She has an M.Phil. from Cambridge University in Computer Speech and Language Processing and a Ph.D. from Stanford University in Linguistics. Her dissertation was a study of conversation in a MUD. She's interested in computer games, science fiction and media fandom.

Elizabeth Reba Weise (left) is national cyberspace reporter for The Associated Press in San Francisco. She also edited *Closer to Home: Bisexuality and Feminism* (Seal Press, 1992). A Seattleite in exile, she still dreams of mossy sidewalks and the scent of Oregon grape in the rain.

Selected Titles from Seal Press

Women's Studies, Lesbian Studies and Popular Culture

Listen Up: Voices from the Next Feminist Generation, edited by Barbara Findlen. $12.95, 1-878067-61-3.

Closer to Home: Bisexuality and Feminism, edited by Elizabeth Reba Weise. $14.95, 1-878067-17-6.

She's A Rebel: The History of Women in Rock & Roll, by Gillian G. Gaar. $16.95, 1-878067-08-7.

The Me in the Mirror by Connie Panzarino. $12.95, 1-878067-45-1.

Lesbian Couples: Creating Healthy Relationships for the '90s, by D. Merilee Clunis and G. Dorsey Green. $12.95, 1-878067-37-0. Also available on audiocassette: 60 minutes, $9.95, 0-931188-85-7.

Egalia's Daughters by Gerd Brantenberg. $11.95, 1-878067-58-3.

The Things That Divide Us: Stories by Women, edited by Faith Conlon, Rachel da Silva and Barbara Wilson. $10.95, 0-931188-32-6.

You Don't Have to Take It!: A Woman's Guide to Confronting Emotional Abuse at Work, by Ginny NiCarthy, Naomi Gottlieb and Sandra Coffman. $14.95, 1-878067-35-4.

Ordering Information

Individuals: If you are unable to obtain a Seal Press title from a bookstore, please order from us directly. Enclose payment with your order and 16.5% of the book total for shipping and handling. Washington residents should add 8.2% sales tax. Checks, MasterCard and Visa accepted. Please keep in mind that we are unable to accept returns.

> Seal Press
> 3131 Western Avenue, Suite 410
> Seattle, Washington 98121
> (206) 283-7844
> (206) 285-9410 FAX
> sealprss@scn.org EMAIL

If ordering with a credit card, don't forget to include your name as it appears on the card, the expiration date and your signature. Please allow us 4 weeks to fulfill your order.